SpringerBriefs in Biochemistry
and Molecular Biology

More information about this series at http://www.springer.com/series/10196

Mahboob Ul Hussain

Connexins: The Gap Junction Proteins

 Springer

Mahboob Ul Hussain
Department of Biotechnology
University of Kashmir
Srinagar, Jammu and Kashmir, India

ISSN 2211-9353   ISSN 2211-9361 (electronic)
ISBN 978-81-322-1918-7   ISBN 978-81-322-1919-4 (eBook)
DOI 10.1007/978-81-322-1919-4
Springer New Delhi Heidelberg New York Dordrecht London

Library of Congress Control Number: 2014942930

Printed on acid-free paper

Springer is part of Springer Science+Business Media (www.springer.com)

Dedicated to
Parents, Showkeya and Rayyan

Preface

The main objective of writing this book on gap junction proteins is to address the audience who are not familiar with the gap junction proteins known as connexins. Connexins, or gap junction proteins, are a family of structurally related transmembrane proteins that assemble to form intercellular channels known as gap junction channels. This book provides a basic overview of the family of connexin proteins. After providing brief introduction about the connexins and their historical perspective, the book furnishes detailed insight about the genomic organization and the expression regulation of the connexin proteins. Moreover, a comprehensive overview of the distribution and role of connexin proteins in various tissues has been addressed. The latter topics have dealt with the role of connexin mutation in various human diseases. I hope my little endeavour will benefit the students who are interested to know the fundamentals of gap junctions. To make this book better, criticisms and suggestions are most welcomed.

Srinagar, Jammu and Kashmir   Mahboob Ul Hussain

Acknowledgment

I would like to acknowledge the help of those people who provided me the logistic and technical support while writing this book. I would like to thank Prof. Rolf Dermietzel (Ph.D. supervisor) for introducing me to the field of connexins. This manuscript would have been impossible without the everlasting support of Prof. Khurshid Andrabi. The technical support of Prof. Georg Zoidl is highly acknowledged. The help of Bilal A Reshi, Assistant Professor in the Department of Biotechnology, is unforgettable. The helpful comments provided by Dr. Mushtaq A Beigh are highly appreciated. I would like to thank my other colleagues at the Department of Biotechnology and to all my teachers. My sincere thanks go to all scientists whose work provided the necessary platform for writing this book.

My sincere thanks go to my parents, my wife Showkeya, and little Rayyan, who provided me the congenial atmosphere at home and were patient enough to allow me to complete this book.

Mahboob Ul Hussain

Contents

About the Author

Dr. M.U. Hussain earned his Ph.D. from International Graduate School of Neuroscience (IGSN), Ruhr University, Bochum, Germany, under the supervision of renowned gap-junction neurobiologist, Prof. Dr. Med Rolf Dermietzel. He worked as a post-doctoral research fellow in the Department of Neuroanatomy and Molecular Brain Research. During his doctoral and post-doctoral work, Dr. Hussain gained research experience in various aspects of modern biology. His doctoral work earned him the highest European honour of "Summa Cum Lauda". To the credit of Dr. Hussain are many international papers, published in well reputed journals. Presently, Dr. Hussain is Senior Assistant Professor at the Department of Biotechnology, University of Kashmir. His main responsibilities in the department are research and post-graduate teaching. In the department, he supervises research scholars and conducts laboratory courses for post-graduate students. Besides, he teaches Molecular Biology, Genetic Engineering and Molecular Techniques to post-graduate students.

Chapter 1
Introduction to Gap Junction

Multicellular organisms with complex tissue systems have evolved over a period from simple unicellular organisms. As opposed to unicellular organisms, which carry out most of their biological processes within a single cell, individual cells within a multicellular organization need to communicate with each other for the successful exchange of nutrients and signals, necessary for the maintenance of the organization. Organisms have evolved multiple strategies to achieve this goal, which include long-range interactions mediated by neural or endocrine mechanisms or short-range interactions that include direct physical or cell–cell contact. This is accomplished in a variety of ways, mostly by the formation of a series of pores, or communicating channels, which can facilitate cell–cell communication. In animal cell system, gap junctions between the cells form one such communication system. The fundamental function of two or more cells, coupled by gap junctions, is clearly to "communicate". While humans communicate with other humans via words, body language, and touch, cells communicate with each other in a multicellular organism via chemical signals. The major physiological role of gap junctions is to synchronize metabolic or electronic signals between cells in a tissue. Cells have four basic functions, namely, (1) to proliferate, (2) to differentiate, (3) to apoptose or die by programmed cell death, and (4) to adaptively respond, if it is already terminally differentiated. In multicellular organism, a delicate coordination or orchestration of these four cellular functions must occur. Growth, differentiation, apoptosis, wound healing, and homeostatic control of differentiated functions must occur in a single space, and this is achieved by coupling the cells within a tissue/organ mainly through gap junctions.

© Springer India 2014
M.U. Hussain, *Connexins: The Gap Junction Proteins*, SpringerBriefs
in Biochemistry and Molecular Biology, DOI 10.1007/978-81-322-1919-4_1

Chapter 2
Historical and Topological Perspective of Connexins

Preceding to the discovery of gap junction, it was observed that there might exist a pathway for direct cell–cell communication. Based on the work of Weidmann, it was observed that the space constant for the spread of current extends beyond the expected value for a single Purkinje fibre. Accordingly, it was suggested that this phenomenon might be due to the existence of a low-resistance intercellular channels. Further evidence supporting the existence of such intercellular channels was provided by the discovery of electric transmission at the giant crayfish motor synapses. These and other observations established that the cells of most invertebrate and vertebrate tissues are directly linked together by communicating channels mediated by low-resistance intercellular channels. In vertebrates, most of the cells form gap junctions except red blood cells, spermatozoa, and skeletal muscle. However, the progenitors of these cells are known to form gap junctions. Direct evidence for the existence of intercellular communicating channels was provided by the electron microscopic studies. It was observed that in excitable tissues, a current transfer between the adjacent cells only occurs when the plasma membrane of these cells was in close proximity and no such electric transmission was detected when the cells were not in close proximity with each other. Further studies confirmed the existence of such intercellular channels, and these were assigned several names, like nexus, macula communicans, and finally gap junction.

The gap junctions' topology has been significantly elucidated using electron microscopy and other biochemical strategies. Transmission electron micrography indicates that the gap junctions assemble in the regions where the plasma membranes of adjacent cells closely approach each other with a small gap of about 2–3.5 nm (Fig. 2.1). Freeze-fracture electron micrography of vertebrate junctions reveals that on the P-face, particles of 8.5 to 9.5 nm exist either singly or in plaque-like arrays with complementary pits on the E-face. Similarly, atomic force microscopy (AFM) indicates a dense packing of particles with a centre-to-centre distance of 9–10 nm. Based on these studies, it was demonstrated that these particles contain a pore-like structure with a diameter of 2 nm, a depth of 1 nm, and a width of 3.8 nm. According to X-ray diffraction studies and Fourier analysis, the gap junction forms a hexagonal

© Springer India 2014
M.U. Hussain, *Connexins: The Gap Junction Proteins*, SpringerBriefs
in Biochemistry and Molecular Biology, DOI 10.1007/978-81-322-1919-4_2

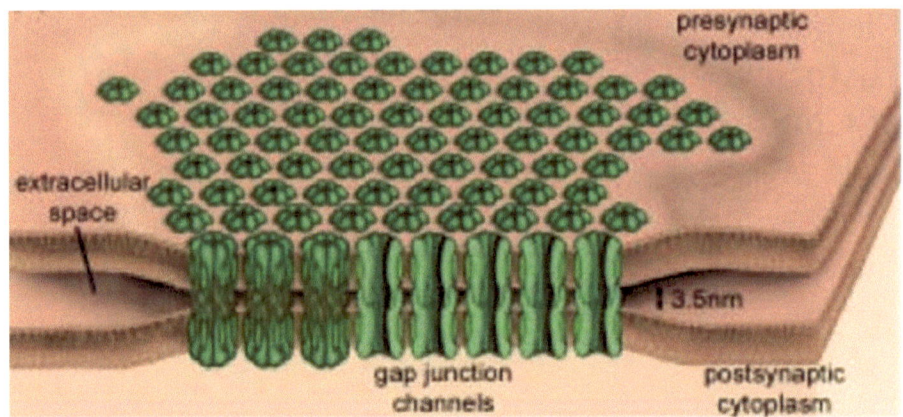

Fig. 2.1 Gap junction channel. Gap junction channels assemble in plaques containing few to several hundred single channels. Each cell contributes one hemichannel called connexon that consists of six connexin proteins. The gap junction channels span a small gap (3.5 nm) between the cell membranes and connect the cytoplasm of neighbouring cells

twisted cylinder with an apparent aqueous pore in the centre. The structure of the gap junction channel has been refined in recent years by using various genetic engineering approaches. Recombinant gap junction proteins and various truncated and mutated versions paved way for elucidating the finer structural details of these channels. Expression and purification of gap junction proteins led to their structural analysis using X-ray crystallographic techniques. In one such study, the X-ray crystallographic structure of a gap junction, at a resolution of 7 A^0, indicates that each hemichannel contains 24 α-helices corresponding to the four transmembrane domains of the six protein subunits.

Chapter 3
Molecular Components and Nomenclature of Gap Junctions

After the confirmation of the existence of gap junction channels, it was imperative to know their molecular composition. In the beginning, it was regarded that all the gap junctions are made of same kind of protein. However, further studies showed that there exist differences in the protein components of the gap junction. For example, the proteins obtained from various gap junction-enriched preparations showed different electrophoretic mobilities in the range of 21–70 kDa, when detected by SDS-PAGE. This concept was further established after performing micro-sequencing of the amino-terminal regions of these proteins that revealed the differences in the primary sequence of the gap junction proteins. Based on the primary sequence information, oligonucleotide probes were synthesized to screen libraries for the existence of other gap junction proteins. Moreover, generation of antibodies also proved instrumental for the isolation of different gap junction proteins. Based on these techniques, a gap junction protein of 32 kDa was isolated from the liver of rat and human cDNA clones. Similarly, a cDNA encoding a related but a different polypeptide of 43 kDa was isolated from rat heart gap junctions. In the following years, many different gap junction proteins were isolated from various cells and tissues. Thus, it became evident that there exists a family of gap junction proteins. Presently we now know that there are about 20 different gap junction proteins existing in the mouse and human genome. After the discovery of different gap junction proteins, their bio-chemical characterization was performed, and it was found that the basic principal component of gap junctions is a membrane protein called connexin (Cx). The gap junctions are assembled from the connexin proteins, and this assembly is hierarchical in nature. Connexins assemble together to form a basic unit of structure called the connexon, which is a hexameric structure with a torrid appearance. An individual connexon from one cell docks or associates with a corresponding connexon on a neighbouring cell to form a gap junction channel. Usually, multiple channels cluster or aggregate in the plane of the membrane to form what is called as gap junction plaques. The question arises whether any other proteins, besides connexins, are part of gap junctions. However, the evidences indicate that the gap junctions are purely made of connexin proteins. For example, reconstitution of purified connexins into

© Springer India 2014
M.U. Hussain, *Connexins: The Gap Junction Proteins*, SpringerBriefs
in Biochemistry and Molecular Biology, DOI 10.1007/978-81-322-1919-4_3

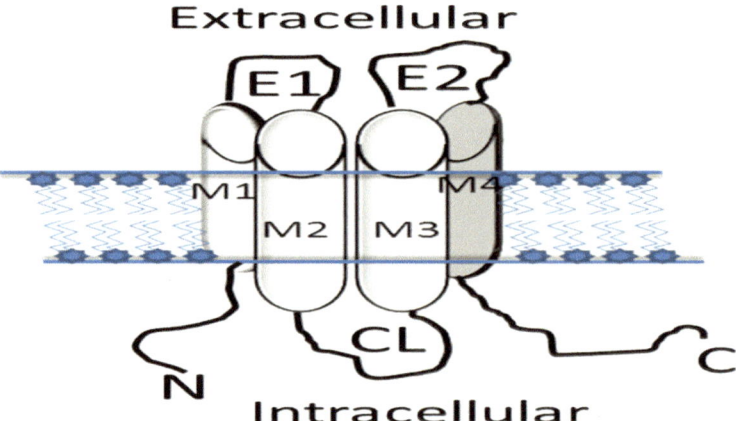

Fig. 3.1 Topological of a connexin protein. The cylinders represent transmembrane domains (*M1–M4*). The loops between the first and the second, as well as the third and fourth, transmembrane domains are predicted to be extracellular (*E1 and E2*); *CL* represents cytoplasmic loop between M2 and M3 transmembrane domains. *N- and C-* represents N-terminal and C-terminal domains

artificial membranes yields functional channels. Moreover, expression of connexin cDNAs in heterologous systems (including yeast) yields not only functional gap junction channels but also gap junctions that are ultrastructurally identical to those occurring naturally in vivo. Further sequence and structural elucidation of connexins demonstrated that each connexin protein is composed of nine main domains. These include four transmembrane domains, intracellular N-terminal and C-terminal domains, two extracellular loops that are stabilized by intramolecular disulfide bonds, and a cytoplasmic loop (Fig. 3.1). The N-terminus, the two extracellular loops, and the four transmembrane domains are highly conserved among different connexin isoforms. In contrast, the cytoplasmic loop and the C-terminal domain are divergent and variable in length and sequence, thus accounting for the functional differences between the different connexins and the connexon types.

Due to the discovery of many different connexins, it was proposed that some nomenclature might be established for naming the connexin proteins. Accordingly, two systems were used to name the connexin proteins: (1) The first is based on a "number" system whereby the molecular weight predicted from the cDNA sequence of the connexin denotes the type of the connexin. The connexin is denoted as "Cx" and then followed by the molecular weight of the protein. For example, Cx26, Cx32, and Cx43 refer to the connexins with molecular weight 26, 32, and 43 kDa, respectively, and (2) the second is based on sequence similarity and length of the cytoplasmic domain of the connexins, thereby classifying them into α-, β-, and γ-subgroups. In this latter system, the connexins are designated as "GJ" for gap junction, followed by a number indicating the order of their discovery within each subgroup. For example, Cx43 was the first connexin discovered in α-subgroup and was designated as Gja1, while Cx32 was the first in the β-subgroup and was designated as Gjb1. In this book, the nomenclature based on the molecular weight of connexin proteins will be followed.

Chapter 4
General Cell Biology of Connexins

4.1 Synthesis, Maturation, and Transport to the Cell Membrane

Similar to other transmembrane proteins, connexin translation occurs on the ribosomes attached to the endoplasmic reticulum (ER), followed by co-translational release of the protein in the lumen of ER. During the transport of connexins from ER to Golgi network, they are folded into the three-dimensional structure, followed by their oligomerization into hexameric hemichannels called connexons. Each connexon subunits can be either homomeric (made of similar connexins) or heteromeric (made of different connexins). The possibility of forming the heteromeric connexin channels is significant, as many cells express more than one type of connexin protein. Hence, more the expression of different connexins in a cell, the possibility of forming various permutation combinations of connexin channels increases significantly. Moreover, channels formed of heteromeric connexons have different properties from those of homomeric channels of the constituent connexins, thereby allowing for critical regulation of the permeability and conductance of gap junctions. However, it is regarded that there exists some selective compatibility between various connexins to form heterotypic channels, and this compatibility has been attributed to the N-terminal and C-terminal domains of each connexin. Furthermore, the transport of connexon hemichannels involves their packaging into vesicles, and then these vesicles deliver the connexon hemichannels to the plasma membrane. The transport of these vesicles to the plasma membrane has been demonstrated to be both microtubule and actin mediated. The insertion of connexon hemichannels to the plasma membrane is regarded to be random. However, some studies have shown that microtubules target the connexon hemichannels to a specific domains or regions of plasma membrane, which are rich in adherens junction proteins. The inserted connexon hemichannels interact with the apposing connexion hemichannels of adjacent cell, thus allowing the formation of complete intercellular gap junction channel. The interaction between the apposing connexon hemichannels is mediated

© Springer India 2014

M.U. Hussain, *Connexins: The Gap Junction Proteins*, SpringerBriefs
in Biochemistry and Molecular Biology, DOI 10.1007/978-81-322-1919-4_4

by the external loop domains. The individual channels aggregate in the membrane to form plaques and hence are known as gap junction plaques. However, under certain conditions, the inserted connexon hemichannels remain uncoupled, thus connecting the intracellular milieu directly to the extracellular. The existence of these hemichannels has been well established and is known to play important roles in cell physiology.

4.2   Post-translational Modifications, Half-Life, and Degradation

Connexins are not regarded as mere intercellular channels that link two communicating cells, but connexins are now regarded to play important role in controlling various other cell functions. In part, these connexin functions are determined by various post-translational modifications, which include phosphorylation, hydroxylation, acetylation, nitrosylation, and palmitoylation. The most important of these connexin post-translational modifications is phosphorylation of various amino acid residues of connexins. Most of these phosphorylation events occur at the C-terminal domain of the connexins, thus modulating the function of each connexins and hence the gap junction channel. Most of the connexins, except Cx26, are known to be phosphorylated, which include Cx31, Cx32, Cx37, Cx40, Cx43, Cx45, Cx46, Cx50, and Cx57. Various kinases have been identified that target connexin proteins, which include Src, PKC, MAPKs, tyrosine kinases, etc. Phosphorylation serves as an important post-translational modification system that regulates various aspects of connexin in including the formation and modulation of gap junction channels. Phosphorylation events regulate the gating of connexin channels and thus control the opening and closing of these channels. The effect of phosphorylation on channel gating is very specific, as the phosphorylation of a connexin isoform such as Cx43 on different residues by the same kinase may lead to opposite effects with respect to enhancing or inhibiting gap junction function or gap junction intercellular communication (GJIC). Besides gating, connexin phosphorylation affects various other aspects of connexins such as trafficking, assembly and disassembly, and degradation. Moreover, connexin phosphorylation at specific residues controls the interaction of connexins with other proteins, and these events affect other cellular functions of connexins that are independent of gap junctions, including the control of growth and proliferation, cell migration, etc.

Just like other proteins in the cell, connexins are degraded in a controlled manner, and their turnover is very fast, with a half-life not exceeding few hours. The newly synthesized connexins are delivered to the membrane and incorporated at the existing gap junction plaques, while the older connexins are internalized from the existing gap junction plaques and finally degraded. Thus, the gap junction plaques are highly dynamic in nature, having newly delivered connexons localized to the periphery of existing gap junctional plaques, while those destined for degradation are present at the centre of the gap junctional plaque. The mechanism of gap junctional internalization is through the formation of annular junctions, which are large double-membrane vesicular structures that could contain the entire gap junction, or a fragment of it, and transport it from cell–cell boundaries into one of the two interacting cells. Following gap junction internalization, these complexes undergo preliminary degradation in the annular junctions, leading to disassembly of gap junctions and connexons into individual connexins. The connexin proteins undergo complete degradation either through proteasomes or through lysosomes, with each of the two degradation pathways having different roles.

Chapter 5
Genomic Organization of Connexins

The general genomic structure of connexin genes is rather simple. The 5′-untranslated region (5′-UTR) of connexin mRNA is part of exon 1 that is separated from exon 2 by a long intron. Exon 2 harbours the entire open reading frame and the subsequent 3′-untranslated region (3′-UTR). However, with the accumulation of new experimental data, it became evident that the genomic organization of several connexin genes refutes this simplicity. The identification of different splice isoforms of several connexin genes demonstrated that exon 1 harbouring the 5′-UTR could be spliced in an alternate manner possibly due to alternate promoter usage. However, it should be emphasized that these transcript isoforms vary only in their 5′-untranslated region, leaving the coding region unaltered. Moreover, there are well-known exceptions to the above-proposed genomic structure. For example, Cx36, Cx39, and Cx57 coding regions are interrupted by another intron. The variation in the genomic organization of various connexin genes can be ascertained by the following paragraphs that describe the structure of several individual connexin genes.

5.1 Cx32

The genomic organization of Cx32 gene consists of two exons, separated by a 6.1 kb intron. Exon1 constitutes the 5′-UTR, while the entire coding region and 3′-UTR are present in exon 2. However, multiple alternatively spliced transcripts of Cx32 gene have been identified in the mammalian species. For example, two different transcripts were identified in the rat and human, while in cow and mice three different transcripts have been identified. Additionally, two promoter elements have been identified in the rat Cx32 gene, and the usage of these promoters is known to be cell type specific. In hepatocytes Cx32 mRNA is transcribed from the promoter 1 (P1) that is located upstream of exon 1, while in Schwann cells it is transcribed from an alternative promoter 2 (P2) that is located upstream of exon 1B. In humans, Cx32 gene consists of three exons, designated as 1, 1B, and 2. Exons 1 and 1B

© Springer India 2014
M.U. Hussain, *Connexins: The Gap Junction Proteins*, SpringerBriefs
in Biochemistry and Molecular Biology, DOI 10.1007/978-81-322-1919-4_5

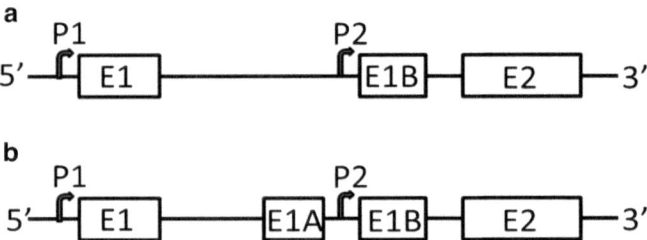

Fig. 5.1 Genomic organization of Cx32: (**a**) human and rat, (**b**) mouse and bovine. "E" represents exons, "P" represents promoter

are alternatively spliced with exon 2 to produce mRNAs with different 5'-UTRs. Transcription initiation occurs from two different sites, and each initiation site is driven by cell type-specific promoter. In liver, the promoter P1 that is located more than 8 kb upstream of the translation start codon is used, while in nerve cells promoter P2, located 497 bp upstream from the translation start codon, is used for the transcription. Mouse Cx32 gene contains at least four exons designated as 1, 1A, 1B, and 2. Three main transcripts have been identified, which arise by the alternative splicing of exon 1, 1A, and 1B with exon 2 (E1/E2, E1A/E2, E1B/E2). The three transcripts are transcribed from two promoters specifically active in different tissues. Transcript E1/E2 (exon 1/exon 2) is transcribed in the hepatocyte from the promoter P1 that is upstream of exon 1, and transcript E1A/E2 is transcribed in embryonic cells and liver from the same promoter P1. However, transcript E1B/E2 is transcribed in the Schwann cell using promoter P2 upstream of exon 2 (Fig. 5.1). The specificity of P1 promoter is determined by the presence of cell type-specific binding site for the hepatocyte nuclear factor-1 (HNF-1). In addition, the promoter P1 contains putative binding site for various ubiquitous transcription factors, such as Sp1/Sp3, nuclear factor 1 (NF-1), and Yin Yang 1 (YY1). Similarly, the specificity of promoter P2 is driven by the presence of binding sites for nerve-specific transcription factors, like Sox10 and early growth response gene-2 (Egr2/Knox20).

5.2   Cx40

Similar to Cx32, various transcripts of Cx40 have been identified in different cell types and in different organisms. Two transcripts are found in the human, three in mouse, and one in the rat. Human Cx40 gene contains at least three exons designated as 1A, 1B, and 2, which are spliced in cell type-specific manner. Transcript E1A/E2 (exon 1A/exon 2) is transcribed in human umbilical cord vein endothelial cells (HUVEC), whereas transcript E1B/E2 (exon 1B/exon 2) is transcribed in human placenta. However, in the human heart, both transcripts have been identified in various regions, like the left atrium, right atrium, left ventricle, and right ventricle.

In mice, the three transcripts arise due to the differential splicing of exon I as exons 1A, 1B, and AS. This results in the formation of three transcript of Cx40 (E1A/E2, E1B/E2, and EAS/E2), having same coding region, however differing in the 5′-UTR. The different transcripts of mouse Cx40 show differential tissue expression. For example, E1A/E2 is ubiquitously expressed, while EAS/E2 is expressed abundantly in the oesophagus. Rat Cx40 is only expressed as single transcript that arises from exon 1 and exon 2 of the Cx40 gene.

5.3   Cx43

Cx43 gene was known to possess two exons, that is, exon 1 containing most of the 5′-UTR and exon 2 containing little part of 5′-UTR and the entire coding region along with the 3′-UTR. However, this general genomic organization of Cx43 has been challenged recently by the identification of many more noncoding exons. It has been demonstrated that, at least in mice, there exist six exons (E1A, E1B, E1C, E1D, E1E, and E2), out of which E1A, E1B, E1C, E1D, and E1E form the 5′-UTR, while E2 contains the coding region and the 3′-UTR. The alternative splicing of these exons results in the production of various transcripts, and these different transcripts are transcribed from three different promoters (P1–P3) (Fig. 5.2). In mice, differential promoter usage and alternative splicing result in the formation of different Cx43 transcripts, which include E1A/E2, E1As/E2, E1A/E1E/E2, E1B/E2, E1Bs/E2, E1Bs/E1D/E2, E1C/E2, and E1C/E1D/E2. In the mouse Cx43 gene, the P1 promoter is located upstream of exon 1A, which was earlier regarded as the main promoter of Cx43 gene. Promoter P2 is located within exon 1A, and promoter P3 is located upstream of exon 1C. Similarly in rat, four exons have been identified (E1A, E1B, E1C, and E2), the alternate splicing of which results in the production of six different Cx43 mRNA transcripts, which includes E1A/E2, E1As/E2, E1A$_L$/E2, E1B/E2, E1Cs/E2, and E1C/E2. In humans, the differential promoter usage and alternate splicing of Cx43 are yet to be identified.

Moreover, Cx43 is the only known connexin, which possesses a pseudogene (ψCx43). The pseudogene is located on human chromosome 5, while the main Cx43 gene is located on the human chromosome 6. The ψCx43 lacks an intron between exon 1 and exon 2, and it has been found that the ψCx43 possesses an open reading frame which is almost similar to that of main Cx43, albeit with some amino acid changes. Interestingly, it has been shown that the Cx43 pseudogene is transcribed specifically in several breast cancer cell lines and not in the normal mammary

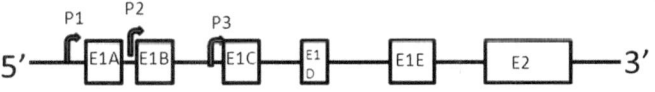

Fig. 5.2 Genomic organization of Cx43: "E" represents exons, "P" represents promoter. E1A–E1E are noncoding exons, while E2 is the coding exon

epithelial cells. There seems to be inverse relation between the expression of Cx43 transcript and ψCx43 transcript. Moreover, it has been demonstrated that the Cx43 pseudogene can be translated in an in vitro translation system and that the Cx43 pseudogene product inhibits cell growth similar to that of Cx43.

5.4   Cx45

Cx45 gene varies a little from the general connexin gene structure. Cx45 gene contains three exons, with 5′-UTR represented by exon 1, exon 2, and part of exon 3. However, the entire coding region and 3′-UTR are present in exon 3. Differential splicing of exon 1 gives rise to various transcripts, differing in 5′-UTR. Accordingly, various alternative spliced transcripts of Cx45 have been identified; each of them shares exon 2 and exon 3, while exon 1A, 1B, and 1C are alternatively spliced with exons 2 and 3. For example, mice Cx45 gene is composed of five exons, designated as exons 1A, 1B, 1C, 2, and 3. Exons 1A, 1B, 1C, and 2 are part of 5′-UTR, while exon 3 contains the remaining 5′-UTR, entire coding sequence, and the 3′-UTR. Additionally, some transcripts have been shown to possess only exon 2 and exon 3. The different transcripts of Cx45 have been shown to arise either due to differential splicing or due to multiple promoter usage. For example, the promoter element, which transcribes the Cx45 mRNA containing exon 1A/2/3, is nearly ubiquitous; the promoter element which transcribes the Cx45 transcript containing exon 1B/2/3 is found in colon, while the transcript containing E1C/2/3 is found in the bladder, lung, skeletal muscle, ovary, and heart.

5.5   Cx26

Cx26 gene possesses the basic genomic organizations, having only two exons, with exon 1 as noncoding and exon 2 as the coding and having 3′-UTR. However, the length of exon 1 and intron (between exon 1 and exon 2) varies among different species. For example, in the mice, the length of exon 1 is 234 bp and that of intron is 3.8 kb. Similarly, in the human Cx26 gene, the length of exon 1 is 160 bp, while that of intron is 3.148 kb. The promoter element of Cx26 gene is known to possess binding sites for various transcription factors, like Sp1/Sp3, with a prominent TATA box.

5.6   Cx31

The Cx31 gene comprises exon 1A, exon 1B, and exon 2. Exons 1A and 1B constitute the 5′-UTR, and exon 2 possesses the coding region and 3′-UTR. However, there are variations from the general structure of Cx31 gene in different organisms.

For example, mice possess the same genomic structure as described above, while in rat, the transcripts with exon 1A and exon 2 have been identified. Mouse 1B is transcribed from a proximal promoter region 561 bp upstream and serves as a basal promoter in both mouse embryonic stem (ES) cells and a mouse keratinocyte-derived cell line. The promoter element of Cx31 does not possess TATA box and contains many GATA-2/GATA-3 binding sites. Moreover, putative binding sites for NFκB, CCAAT-box, cEBPα/CEBβ, cAMP response element, and multiple E-box/E-box are also known to be present in the promoter region.

5.7   Cx30

Gene organization of Cx30 consists of at least six exons, named as exons 1, 2, 3, 4, 5, and 6. Among these, exon 1 to exon 5 constitute the 5′ UTR, while exon 6 contains the coding information and the 3′-UTR. Differential splicing of exon 1 to exon 5 results in the formation of various transcripts of Cx30. In humans, all six exons have been identified with tissue-specific differential expression. For example, in hair follicle keratinocytes, four different transcripts are known to express, and these transcripts share only exon 5 and exon 6. The transcripts of Cx30 are transcribed from a TATA box-containing promoter element, flanking upstream of exon 1. The promoter element contains putative binding sites for Sp1 and early growth gene (Egr) transcription factor.

5.8   Cx36, Cx39, and Cx57 Genes

Exceptions to the general genomic organization of connexins are the Cx36, Cx39, and Cx57 genes. The coding region of these connexin genes is interrupted by introns. The coding regions of Cx36, Cx39, and Cx57 genes are located on two (or more) different exons. In rat Cx36, the coding region is interrupted by a 1.14 kb intron at the N-terminal region, separating the first 71 bp from the rest of the coding region. Similarly, the coding region of the mouse Cx39 is also interrupted by a 1.5 kb intron. In mouse retinal Cx57 gene, a part of C-terminal domain is spliced with another exon 3. As a result, 97.6 % (480 amino acids) of the coding region of mouse Cx57 gene is located on exon 2, whereas the residual 2.4 % (12 amino acids) is encoded by the third exon, which is separated by an intron of about 4 kb. The connexin genes whose coding region is interrupted by introns, the exons are required to be spliced properly for the proper translation. Otherwise, alternative splicing would lead to a dramatic modification of the connexin coding region.

Chapter 6
Transcriptional Regulation of the Connexin Gene

As described previously, alternate splicing and differential promoter usage can give rise to various transcripts of a particular connexin gene. Some of these transcripts are ubiquitously expressed, while some are tissue specific. The transcription of connexin genes is controlled by of various transcription factors and other biological substances. The transcription factors play an important role in regulating the expression of connexin gene. The types of transcription factor(s) determine whether a particular connexin is expressed ubiquitously or in cell type-specific manner. In the following sections, the role of various transcription factors and other signalling molecules that regulate the ubiquitous or tissue-specific expression of connexin genes will be discussed.

6.1 Regulation by Ubiquitous Factors and Signal Transduction Pathways

The basal level of connexin transcription is established by the RNA Pol II promoter elements and the basal transcription factors. The ubiquitous expression of connexins is determined by the various housekeeping transcription factors. Some of these transcription factors are discussed below.

6.1.1 Sp1 and Activator Protein 1 (AP-1)

The Sp family of transcription factors constitute the ubiquitously expressed proteins. Sp1 is an important member of Sp family of transcription factors that recognizes GCGG sequences, called as GC boxes. DNA-binding domain of Sp1 consists of three zinc fingers, which recognize and bind the GC box. Sp1 regulates the basal

© Springer India 2014   17
M.U. Hussain, *Connexins: The Gap Junction Proteins*, SpringerBriefs
in Biochemistry and Molecular Biology, DOI 10.1007/978-81-322-1919-4_6

transcription of several connexin genes, such as Cx32, Cx40, Cx43, and Cx26. In addition to Sp1, Sp3 is known to bind the P1 promoter elements and control the expression of rat Cx40 and Cx43. The cumulative effect of Sp1 and Sp3 binding is required for the transcription of various connexin genes. The P1 promoter element of Cx32 possesses two binding sites for Sp1 transcription factor, and the binding regulates only its basal transcription and not the tissue-specific expression. The binding of Sp1 and Sp3 to the promoter elements of connexin gene is regulated by various signalling molecules and signalling pathways. Moreover, during certain pathophysiological conditions, the Sp1/Sp3 transcription factors are known to be involved in the deregulation of connexin expression. For example, high blood glucose concentration downregulates Cx40 expression in the vascular endothelial cells and that results in the impaired endothelial capillary network formation. The decreased expression of Cx40 has been attributed due to high glucose-induced downregulation of Sp1 transcription factor.

Activator protein 1(AP1) is a basal transcription factor, which activates the transcription of its target genes. AP1 acts as dimer and consists of either Jun proteins (Jun homodimer) or Fos proteins (Fos homodimer) or the combination of Jun with Fos (Jun–Fos heterodimer) or even heterodimer of these proteins with other proteins. AP1 transcription factor forms a characteristic DNA-binding domain called leucine zipper. AP-1 binds to a consensus sequence [TGAC (T/G) TCA] in the promoter and induces transcription of the gene. One or more AP-1-binding sites have been identified in the mouse, rat, and human Cx43 proximal promoters P1. The rat Cx43 promoter contains two AP-1-binding sites, whereas the mouse and human promoters each have one AP-1-binding site. Although the AP1 regulates the basal transcription of various connexin genes, however, under certain circumstances, it induces the expression of certain connexin genes. For example, during the onset of labour, AP-1 transcription factor is necessary for the upregulation of Cx43 expression.

The AP-1 transcription factors are the target of various signalling molecules and pathways that in turn results in the regulation of connexin expression. For example, stimulation of urinary bladder smooth muscle cells by the basic fibroblast growth factor (bFGF) upregulates Cx43 transcription via extracellular signal-regulated kinase (ERK) 1/2-AP-1 pathway. The binding of AP1 to the Cx43 promoter is regarded essential for the upregulation of Cx43 transcription upon stimulation with bFGF. Similarly, TGF-beta induces Cx43 gene expression in normal murine mammary gland epithelial cells. The signalling pathway that activates c-Jun/AP-1 is essential for mediating TGF-beta1-induced Cx43 gene expression.

6.1.2 Cyclic AMP

Cyclic AMP (cAMP) is a ubiquitous secondary messenger, which is formed in response to some extracellular stimuli that activate G protein-coupled adenylate cyclase. The cAMP in turn activates protein kinase A (PKA) and that mediates various functions of cAMP by phosphorylating various target proteins. The adenylate cyclise converts ATP to cAMP at the inner surface of plasma membrane. The cAMP modulates the expression of various genes by binding and thus activating certain proteins known as cAMP response element binding (CREB) proteins. These proteins recognize certain sequence elements in the DNA, known as cAMP response elements (CRE). The promoter elements of various connexins contain cAMP response elements and thus are target of cAMP-mediated regulation. Cx43 promoter element I contains putative cAMP response element, and the expression of Cx43 is enhanced when cells are treated with the cAMP analogue 8Br-cAMP. The signalling mediated by cAMP employs short-term and long-term effects on the regulation of Cx43. The short-term effects include the increased redistribution of Cx43 to the cell membrane, and the long-term effects include increased transcription of Cx43 gene.

6.1.3   Retinoids

Like cAMP, retinoids are known to induce gap junctional intercellular communication and upregulate Cx43 expression in normal and preneoplastic cells. Treatment with the synthetic retinoid tetrahydrotetramethylnapthalenylpropenyl-benzoic acid (TTNPB) is known to cause an approximate tenfold elevation of Cx43 gene transcripts. However, the molecular mechanism for the upregulated expression of Cx43 remained to be understood. It has been hypothesized that retinoic acid increased Cx43 expression in cells by increasing transcription and by stabilizing Cx43 transcripts. The role of retinoic acid on the expression of Cx43 has been studied with reference to human endometrial stromal cells. It has been shown that the retinoic acid upregulated Cx43 mRNA and protein expression with the concomitant increase in the gap junction intercellular communication in the human endometrial stromal cells. The endometrial stromal cells express retinoic acid specific nuclear receptors and the binding of retinoic acid to these receptors results in the upregulation of the transcription of Cx43 gene. In addition to increasing Cx43 mRNA and protein levels, retinoic acid also enhances the dephosphorylation of Cx43, so that the nonphosphorylated form increases with respective to phosphorylated forms. The decrease in the phosphorylation status of Cx43 increases the gap junction communication between the cells.

Similar studies in human skin epidermis have shed more light on the regulation of connexin expression by retinoic acid. Human keratinocytes are connected by many gap junctions, and these junctions are mostly made from Cx26 and Cx43 proteins. It is well known that connexins are involved in keratinocyte differentiation and retinoic acid provides the necessary link by regulating the expression of connexins during cell differentiation. The mechanism by which retinoic acid modulates the differentiation of the epidermis is still not well understood. However, it is regarded that retinoic acid changes the expression of connexins and modulates the gap junction mediated cell–cell communication.

6.1.4   Wnt Pathway

Wnt genes encode a large family of secreted glycoproteins, which play important roles in directing cell fate and cell behaviour and in tumorigenesis. Wnt signalling pathway results in the activation of certain transcription factors, like TCF/LEF proteins. Nuclear localized β-catenin/TCF complexes regulate the transcription of target genes by binding to the consensus sequence A/TA/TCAAAG, known as the TCF/LEF binding site. Various studies have demonstrated the co-localization of Wnt and connexin expression. Connexins are known to have an important role in regulating cell growth and differentiation. Connexin genes are regarded as the targets of Wnt signalling. Interestingly, rat Cx43 promoter P1 contains two TCF/LEF binding consensus sequences in opposite orientations. Two similar TCF/LEF motifs are also found in the human and mouse Cx43 promoters. One of the prime targets of Wnt signalling is the expression regulation of Cx43 in cardiomyocytes. Wnt signalling and Cx43 expression have been shown to be important for the normal development of the heart. Besides regulating expression of Cx43, several lines of evidence suggest that the Wnt signalling also modulates the gap junction channel activity of Cx43. The molecular mechanisms involved in the Wnt signalling and Cx43 are still emerging. One such mechanism involves the Wnt-dependent accumulation of β-catenin near the membrane and the transcriptional activation via a β-catenin/TCF nuclear complex. Since Cx43 is known to interact with β-catenin, the activation of Wnt signalling results in the enhanced interaction between the Cx43 and the β-catenin as part of a complex within the junctional membrane. The sequestration of β-catenin by Cx43 serves to negatively regulate its own transcription, as well as other β-catenin/TCF-dependent transcriptional targets. These findings are corroborated by the observations demonstrating the upregulation of β-catenin levels in various human cancers that act as an important oncogenic signal for the neoplastic transformation of the cells. Since Cx43 is known for its tumour suppressor activity, hence its sequestration by increased β-catenin levels may serve an important event for deregulating the growth of the cells. In addition to the role in regulating cell growth, Wnt–β-catenin–Cx43 signalling is also involved in human cardiomyopathies. During cardiomyopathies, expression of Cx43 is either downregulated or abnormally localized in the cell. The abnormal expression of Cx43 has been linked with decreased β-catenin activation of Cx43 promoter. These observations indicate that the alterations in Wnt signalling influence β-catenin function, which in turn influence gap junction channel gene expression in the heart. These facts are confirmed by the downregulation of β-catenin and concomitant decrease in the Cx43 abundance during various forms of heart disease.

6.2   Regulation by Tissue-Specific Factors

The expression of some connexins shows differential spatial and temporal expression. These connexins are only expressed in certain cell types, and many connexins show development regulation. Besides ubiquitous transcription factors, tissue-specific expression is mediated by additional cell type-specific transcription factors and the signalling pathways that activates such factors. Some of the cell type-specific transcription factors and the signalling molecules involved in tissue-specific expression of connexins are discussed below.

6.2.1   Homeodomain Transcription Factors

Homeodomain-containing proteins are the family of transcription factors that regulate gene expression during development. The homeodomain is 60 amino acid motif that recognizes a specific stretch of 180 bp DNA sequence in the promoter region of various genes, known as homeobox. Nkx2.5 is a transcription factor, which is expressed mostly in cardiac tissue and is critical for the proper development of heart. This protein factor belongs to the family of homeodomain containing transcription factors. Nkx2.5 shows a wide range of expression in the cardiac tissue from drosophila to humans. Cx43 is the main connexin that is expressed in cardiac tissue, and the promoter element P1 of Cx43 contains consensus binding site for Nkx2.5 transcription factor. Nkx2.5 activates the expression of Cx43 and under certain conditions may repress it. Moreover, Nkx2.5 is also known to bind the promoter element of another cardiac connexin, Cx40, and activates its expression. In addition, some studies have pointed out the role of Nkx2.5 in the regulation of cardiac-specific Cx45. Another homeodomain transcription factor that is known to influence the transcription of connexins is Shox2. Shox2 binds the promoters of Cx40 and Cx43 and activates their transcription. In mouse cardiac cells, Cx40 transcription is further regulated by homeodomain-only protein (Hop). The Hop encodes a 73 amino acid protein that contains a domain (the 60-amino acid homeodomain) homologous to those seen in homeobox transcription factors. Unlike all other known homeobox transcription factors, Hop does not directly bind DNA. It is expressed in the embryonic heart and plays an important role in development of the adult cardiac conduction system. Various studies have demonstrated that the expression of Hop is required for the proper expression and localization of Cx40 in the cardiac conduction system.

6.2.2   T-Box Transcription Factors (Tbx5, Tbx2, Tbx3)

T-box transcription factors form a large family of proteins that possess a helix–loop–helix type DNA-binding domain. They are regarded critical for various developmental fates. This family of proteins possesses both transcriptional activators (Tbx5) and transcriptional repressors (Tbx2 and Tbx3). The helix–loop–helix domain of these proteins recognizes the same sequence on the DNA, and hence both activators and repressors bind the same sequence. Thus, the outcome of the binding of T-box transcription factors on the gene expression is determined by the relative concentration of activators and repressors. Several T-box binding sites have been identified in rat and mouse Cx40 and Cx43 promoters. The T-box protein expression coincides with the expression of Cx40. Tbx5 is a T-box transcription factors that is known to bind at multiple sites on the Cx40 promoter. Tbx5 haplo-insufficent mice show marked decrease in the Cx40 mRNA. Besides Tbx5, two other T-box transcription factors are known to influence the expression of Cx40. These include Tbx-2 and Tbx-3 that are known to act as transcriptional repressors. These facts are based on the observations that there exists an inverse relation between the expression of Tbx-2 and Cx40 mRNA in various regions of the developing heart. For example, during mouse embryonic stages 9.5–14.5, the presence of Tbx2 in the cardiac chamber of the embryos is inversely related with the expression of Cx40 and Cx43. Similar to Tbx2, the Tbx3 shows an inverse relation with the expression of Cx43 and Cx40 during cardiac development. The Tbx3 specifically interacts with the TBE-containing DNA region of the Cx43 promoter, indicating that Tbx3 directly represses Cx43.

6.2.3   GATA Family

GATA transcription factors belong to the category of transcription factors that possess two zinc finger DNA-binding domains. They recognize the DNA with a consensus sequence of (A/T) GATA (A/G). Based on their expression profile, the GATA proteins have been divided into two subfamilies, GATA-1, GATA-2, and GATA-3 and GATA-4, GATA-5, and GATA-6. The GATA-1, GATA-2, and GATA-3 genes are prominently expressed in haematopoietic stem cells, where they regulate differentiation and specific gene expression of T-lymphocytes, erythroid cells, and megakaryocytes. GATA-4, GATA-5, and GATA-6 genes are expressed in various mesoderm- and endoderm-derived tissues such as the heart, liver, lung, gonad, and gut, where they play critical roles in regulating tissue-specific gene expression. GATA transcription factors specifically regulate the expression of connexins in various tissues. Rat Cx40 promoter contains putative binding sites for GATA-4, and it has been shown that GATA-4 expression induces the transcription of Cx40.

6.2.4   HNF-1

HNF-1 is a liver-specific transcription factor, which regulates the expression of many liver-specific genes. Two isoforms, HNF-1α and HNF-1β, are known, and these show developmental expression regulation. The HNF-1α is highly expressed in adult liver, whereas the HNF-1β is expressed during the embryonic development. The HNF-1 functions either as a homodimer or as a heterodimer. The promoter region of the mouse Cx32 gene contains two putative binding sites for HNF-1. HNF-1 is mainly responsible for the cell-specific expression of Cx32. These observations are based on the fact that the HNF-1α null mice show significantly decreased Cx32 expression.

6.2.5 Oestrogen

Oestrogen is a steroid hormone, which directly influences the expression of many genes by binding and subsequent activation of oestrogen receptor. Oestrogen receptors constitute the widely studied nuclear receptor. There are two subtypes of the oestrogen receptor, which are known as ER-α and ER-β, and they are the products of two different genes and show tissue-specific expression. In addition to the nuclear receptor-mediated action, oestrogen also influences various physiological processes by influencing various other signalling pathways. Oestrogen influences the expression of connexins, in particularly the Cx43. As described above, oestrogen exerts its effect by binding to nuclear receptors, and the complex binds specific DNA sequences in the promoter region of target genes, called as oestrogen response elements. The role of oestrogen on the connexins transcription has been studied with reference to the Cx43 upregulation during the onset of labour. It is well established that the Cx43 plays a critical role during the onset of labour by increasing the myometrial cell coupling and their coordinated synchronous contraction. This phenotypic effect is strongly correlated with the dramatic increase in both mRNA and protein levels of Cx43 in the myometrium. Interestingly, rat Cx43 promoter has been shown to possess oestrogen response elements, and these elements are responsive to the oestrogens, thus responsible for the increase of Cx43 expression.

6.2.6 Thyroid Hormone

The role of thyroid hormone in regulating the expression of connexins has been studied with reference to Cx43. The role of thyroid hormone in regulating Cx43 transcription was deduced from the observation that the treatment of rat liver samples with thyroid hormone resulted in elevated Cx43 mRNA. Subsequent sequence analysis of the Cx43 upstream promoter element leads to the discovery of thyroid hormone response elements located at positions −480 to −464. Additionally, it is regarded that the −480 position in the rat connexin43 forms stronger complexes with thyroid hormone receptor alpha/retinoid X receptor alpha heterodimers than with vitamin D receptor/retinoid X receptor alpha heterodimers. The physiological relevance of the thyroid hormone in regulating connexin expression has been studied in Sertoli cell of testis. It is well known the thyroid hormones regulate Sertoli cell proliferation and differentiation in the neonatal testis. Further studies have shown that thyroid hormone inhibits Sertoli cell growth and proliferation by inducing the expression of Cx43. The upregulation of Cx43 expression was associated with increased gap junction communication between the cells. These studies were further corroborated by the inhibitors of gap junction coupling that were shown to significantly reverse the inhibitory effect of thyroid hormone on the Sertoli cell growth and proliferation.

The importance of thyroid hormone regulation of connexins expression is evident from the association of hypothyroidism with the expression of various connexins in the thyroid cells. The outcome of iodine deficiency-related hypothyroidism is the decreased gap junction communication between the thyroid cells. Similarly, in autoimmune-related hypothyroidism, the thyroid cells show reduced expression of Cx43, Cx32, and Cx26. Moreover, the thyroid-stimulating hormone (TSH), produced in the hypothalamic region of the brain, regulates thyroid hormone production by increasing the connexin function of thyroid cells. Hence, thyroid hormone acts as an important signalling molecule that regulates the expression of various connexins. Moreover, the expression of connexins is regarded essential for the proper release of thyroid hormone from the thyroid gland.

Chapter 7
Epigenetic Regulation of Connexins

Besides regulation of connexin expression by the direct binding of transcription factors to their regulatory elements, various other epigenetic mechanisms are involved in the expression of connexin gene. Regulation at the epigenetic level is a prerequisite for other regulatory mechanisms to function. Epigenetic processes play an important role in the regulation of connexin gene expression. This type of regulation has important physiological significance, and any disruption in this type of regulation has severe consequences on the cell physiology. One such example is the role of epigenetic factors on the cell growth and proliferation. Connexins are known to regulate cell proliferation, and it is well established that they possess tumour suppressor function. Thus, it is anticipated that the connexins are the target of various epigenetic regulatory mechanisms. Involvement of various epigenetic mechanisms in connexin expression and function is discussed below.

7.1 Histone Acetylation/Deacetylation

Like other eukaryotic genes, connexin genes are present on various chromosomes and thus are warped in protein–DNA complex. Histones form the important protein components of these complexes. Histones are subject to post-translational modification by enzymes, and these modifications include methylation, citrullination, acetylation, phosphorylation, SUMOylation, ubiquitination, and ADP-ribosylation. The main function of these modifications is to affect the gene expression by remodelling the chromatin structure of a particular gene. For the transcription of connexin genes, the chromatin structure needs to be modified, and histones are the prime targets of these modifications. Studies have demonstrated that the acetylation/deacetylation constitutes an important mechanism of the connexin gene regulation. For example, the use of various histone deacetylase inhibitor results in the increased Cx43 transcription. The increased transcription of Cx43 is attributed due to more acetylation of H3 and H4 histones of the chromatin near the promoter region of Cx43. In one such study,

© Springer India 2014
M.U. Hussain, *Connexins: The Gap Junction Proteins*, SpringerBriefs
in Biochemistry and Molecular Biology, DOI 10.1007/978-81-322-1919-4_7

it was shown that the treatment of prostate cancer cells with trichostatin A (TSA) results in the increased acetylation of histones in the Cx43 promoter. The increased acetylation was found due to the recruitment of p300/CREB-binding protein, a transcriptional coactivator displaying histone acetyltransferase (HAT) activity, to the promoter region of Cx43. Because of the chromatin remodelling, the binding of transcription factors AP-1 and Sp1 to the Cx43 gene promoter is enhanced manifold with the concomitant increase in Cx43 transcription.

7.2   DNA Methylation

DNA methylation is one of the important epigenetic mechanisms of controlling the expression of genes. DNA methylation dynamics are mostly associated with the promoter regions of genes. Promoter methylation occurs mostly at the CCGG sequences known as CpG islands. In this context, hypermethylation of promoter regions results in the silencing of the gene expression. DNA methylation of promoter regions of connexins has been well studied, and most of these are frequently studied in a clinical context. Downregulation of connexin expression has been associated with the hypermethylation of their promoter. Cx26, Cx32, and Cx43 promoter elements have been found hypermethylated in various malignant tissues like cultured human lung cancer cells, primary human renal carcinoma cells, cultured human oesophageal cancer cells, etc. Similarly, in liver carcinogenesis, Cx26 expression is downregulated, and this has been found to have strong correlation with the hypermethylation state of its promoter. The mechanism responsible for the connexin gene silencing and hypermethylation is not well understood. Many factors have been attributed as a link between DNA hypermethylation and impairment of connexin expression. It has been shown that hypermethylation results in the aberrant binding or recruitment of transcription factors to the promoter elements. For example, Cx43 promoter has been found hypermethylated in human primary non-small cell lung carcinoma, and this has been related with the reduced binding of AP1 transcription factor. Moreover, the Sp1 binding site overlaps with the CpG islands, which are the prime targets of methylation. In addition, Sp1 is a ubiquitous transcription factor involved in the expression of various connexin genes. Thus, it may be construed that hypermethylation will result in the aberrant binding of Sp1 transcription factor. In fact, Sp1 binding sites of the Cx26 gene promoter and the Cx32 gene promoter in cultured and primary human breast cancer cells have been found hypermethylated with the concomitant decrease in the expression of these connexins.

Chapter 8
Translational Regulation of Connexins

The transcriptional control constitutes a major step in the regulation of gene expression. However, proteins are the major cell entities that act as the actual business ends of the cell physiology. The transcriptional regulation of gene expression is just an initial step in the formation of proteins. Translation of mRNAs into proteins constitutes another important step in regulating the expression of proteins. For the proper functioning of a cell, the functional entities, which include the proteins, must be present at a defined concentration, at a particular time, and at a proper location. At a particular time, cells possess thousands of mRNA molecules, each competing for the components of translational machinery. Thus to avoid such competition, it becomes imperative for a cell to regulate the translation of its different mRNAs. Moreover, cells are always exposed to varying conditions, and for its survival, it needs to respond properly and in a particular period. Regulation at the transcriptional level takes some time, while regulation at the translational level is a rapid and more direct one. Thus, translational regulation gives a cell the required adaptability to respond to the changing conditions. Translational regulation can occur at multiple steps, like initiation, elongation, and degradation. Regulation at the initiation step is the most critical one, and this allows for the precise spatial and temporal fine-tuning of protein levels to permit normal physiological function. The regulation of connexin translation can be studied at two levels. One is cap-dependent and another is cap-independent translation. The role of these two means of translational regulation of connexin expression is discussed below.

8.1 Cap-Dependent and Cap-Independent Translation

Translation initiation in eukaryotic mRNAs is a highly regulated process that accounts for the last step of gene expression control. For a majority of the eukaryotic mRNAs, the ribosome associates with the mRNA by virtue of the cap structure, a 7-methyl guanylic acid residue at the 5′ terminus. Then, this cap-binding complex

© Springer India 2014
M.U. Hussain, *Connexins: The Gap Junction Proteins*, SpringerBriefs
in Biochemistry and Molecular Biology, DOI 10.1007/978-81-322-1919-4_8

scans the 5′-UTR till it finds the start codon. While most mRNAs initiate translation by the above-discussed mechanism, a growing number of mRNAs appear to follow different rules, wherein certain *cis* elements present in the mRNA were found, enough to recruit the translational machinery without the need of a cap structure, hence named as cap-independent translation. These cis-acting are termed as the internal ribosome entry site (IRES).

Translation of connexins is usually mediated by a classical cap-dependent mechanism, where an initiation complex is assembled at the 5′ cap of the connexin mRNA. The initiation complex scans the 5′-UTR of connexin mRNAs until an authentic start codon is reached. Besides the classical cap-dependent translation, connexin mRNAs are known to be regulated by cap-independent translational mechanisms. The importance of exon 1 in the expression of a connexin was recently observed in the Cx32. A point mutation in the 5′-UTR of the Cx32 gene was found in patients with the X-linked form of Charcot–Marie–Tooth disease (CMTX), a neurological disorder characterized by demyelination of the peripheral nerves and leading to paralysis at an early age. When this mutation was tested in transgenic mice, the mRNA synthesis and splicing of pre-mRNA were found to occur normally, but translation was completely blocked. This observation led to the discovery of an internal ribosome entry site (IRES) in the 5′-UTR of Cx32 mRNA. IRES-mediated initiation of translation was first discovered in picornavirus and later also in some other viral RNAs that do not have a cap structure at their 5′ end. More recently, several cellular genes have also been shown to contain IRES elements. In most cases, however, it is not known whether the IRES elements are essential for expression of those genes. The fact that a mutation in the IRES of Cx32 causes a CMTX phenotype indicates that the IRES are essential for the expression of that gene in peripheral nerves, specifically, in Schwann cells. This represents one of the first examples of an essential IRES element in the expression of a normal cellular gene and of an IRES mutation causing a human disease. An even more active IRES element has since been discovered in the 5′-UTR of the Cx43 gene. A similar type of internal ribosome entry site was found in Cx26. The IRES in Cx43 and Cx26 are known to play an important role in the translation of these connexins during contact inhibition. Interestingly, an internal IRES element (in the coding region of exon 2) was found in the zebrafish connexin (ZfCx55.5), which results in the separate translation of carboxy-terminal domain of this connexin. This mechanism has added a new dimension in the translation of connexins, which can explain some of the functions of connexins that are regarded as gap junction independent. Most cellular mRNAs, presumably translated from the cap structures at their 5′-ends, have relatively short 5′-UTRs, <50 nucleotides long. Cap-mediated translation of such mRNAs is efficient because ribosomes traverse only a short distance between the site of attachment to the mRNA (cap structure) and the translational initiation site. In contrast, the 5′-UTRs of mRNAs translated from IRES elements typically are considerably longer, which is very inefficient for cap-mediated translation. Comparison of the 5′-UTRs of various connexin genes revealed that most of the 5′-UTRs of connexins are all much longer than 50 nucleotides. If the other connexin genes indeed also contain functional IRES elements, then it is interesting to know why connexin genes have acquired IRES elements for their efficient translation. One speculation is that connexins have a very important cellular function that requires that these proteins to be translated even under the conditions when global translation is inhibited. Indeed, IRES-mediated initiation of translation has been found in genes of proteins that need to be expressed under conditions where cap-mediated translation is impaired. This includes some transcription factors, heat shock proteins, and proteins involved in apoptosis and cell cycle regulation. Cap-mediated translation initiation is typically inhibited during mitosis, apoptosis, and conditions of cellular stress. Connexins may be an additional family of proteins for which their continued synthesis must be ensured under conditions where cap-mediated translation is impaired. A further advantage of IRES-mediated translation may be the rapid initiation of translation of pre-existing connexin mRNA in response to cellular signals without affecting the rate of overall protein synthesis. One of the theories of the origin of cancer proposes that most cancers arise from mutations in stem cells rather than mutations in differentiated cells. Because of their stem cell-like properties, malignant tumour cells have lost their ability to communicate with their neighbours and thus are unable to maintain their differentiated state. The importance of cellular communication in preventing dedifferentiation is illustrated by the fact that reintroduction of functional connexin genes into a cancer cell phenotypically reverts the cell to normal. Tumour promoters, for example, phorbol esters, are known inhibitors of gap junction communication. In their presence, a mutated initiated stem cell loses its ability to differentiate and remains in, or returns to, the immortal state of a stem cell, which because of the mutation now proliferates unchecked. Thus, the expression of connexin genes must be allowed under all physiological conditions, even when cap-mediated translation is suppressed. This would include mitosis, apoptosis, and cellular stress situations, such as heat shock. Evidence indicates that gap junctional communication is required for apoptosis to occur. Cancer cells typically cannot apoptose because of the absence of gap junctional communication. To ensure continued gap junctional communication, rapid initiation of translation of pre-existing connexin mRNA must be possible in response to specific cellular signals. The presence of IRES elements in the 5′-UTR of gap junction genes would satisfy this requirement. It would also explain why during evolution all connexin genes have acquired the additional exon upstream of the coding exon. Whether the first exons of other connexin genes also contain IRES elements, as would be predicted by this hypothesis, remains to be seen.

8.2 MicroRNAs (miRNAs) and Translational Regulation

MicroRNAs (miRNAs) are an endogenous class of small noncoding RNAs, typically 18–22 nucleotides in length that regulate gene expression by binding to the 3′-UTR of the target mRNAs. Their conservation in different species is suggestive of their physiological importance and an indication of their participation in the essential biological processes. Except for the rapidly evolving RNA Pol III transcribed miRNA clusters, all the miRNA genes are transcribed by RNA Pol II in the nucleus. The primary transcript of a microRNA gene is called the primary miRNA (pri-miRNA), and in most of the cases, it possesses a single 70-nucleotide-long hairpin, a 5′ cap, and a 3′ poly (A) tail. The pri-miRNAs are processed by the nuclear processor complex (Drosha and its cofactor DGCR8) into ~60-nucleotide precursor miRNA called as the precursor miRNA (pre-miRNAs). The pre-miRNAs are transported into the cytoplasm by the Ran-GTP-dependent exportin-5 transporter. In the cytoplasm, the pre-miRNAs are cleaved into ~21–25 nucleotide mature miRNAs by the riboendonuclease called Dicer. In the final step of miRNA biogenesis, one of the strands of the 22-nucleotide miRNA duplex is incorporated into a ribonucleoprotein complex, called miRNA-induced silencing complex (miRISC). The miRISC targets the specific mRNA and the specificity of the target selection is determined by the complementarity between the seed region (2–8 nucleotides from 5′ end) of the miRNA and its target site in the 3′-UTR of the mRNA. The binding of the miRISC to the target mRNA results either in the mRNA cleavage or the translational repression.

Growing evidence shows that microRNAs (miRNAs) regulate various developmental and homeostatic events in vertebrates and invertebrates. Recently, miRNAs have emerged as important regulators in various developmental, physiological, and pathological conditions such as tumorigenesis, viral infection, and cell differentiation.

The role of miRNAs in connexin expression regulation has been recently investigated, and the data indicates that connexins are the candidate genes for the miRNA-mediated regulation. The role of miRNAs in regulating connexin expression has been studied with reference to Cx43. miRNA analysis indicated that Cx43 is the major target of various miRNAs as compared with the other gap junction proteins. The interplay of miRNAs and connexin regulation has important physiological and pathophysiological significance. This was first demonstrated in osteoblast differentiation. Expression of Cx43 is regarded critical for the proper differentiation of osteoblasts and bone formation. During the bone differentiation process, the expression of Cx43 is increased, and this upregulation has been correlated with the concomitant decrease in the miR-206. Subsequent studies have shown that the miR-206 binds the 3′-UTR of Cx43 mRNA and brings about the translational downregulation. These observations were further confirmed by the over-expression of miR-206 in osteoblasts that results in the suppression of their differentiation. Moreover, knockdown of miR-206 in the osteoblasts depicted enhanced cellular differentiation. Interestingly, transgenic mice expressing miR-206 in osteoblasts possessed underdeveloped bone formation with low bone mass. Furthermore, it has been demonstrated that during the differentiation of C2C12 myoblast cells into multinucleate myotubes, miR-206, miR-1, and miR-133 play important roles by decreasing the levels of Cx43. It has been demonstrated that the miR-206 and miR-1 can bind the 3′-UTR of Cx43 and downregulate its translation.

Besides their involvement in various physiological processes, many disease conditions have further highlighted the importance of miRNAs in regulating connexin expression. For example, it is well known that the expression of Cx43 is substantially increased in response to hypertrophic stress in cardiomyocytes. The up- regulation of Cx43 has been significantly correlated with the downregulation of miR-1 in the hypertrophic heart samples. In fact, it has been established that the binding of miR-1 and the downregulation of miR-1 leads to increased translation of Cx43 mRNA. Besides being the direct target of miRNAs, the connexin-formed gap junction channels have been demonstrated to constitute an important route for the intercellular trafficking of miRNAs. Hence, gap junctional communication provides an important route for the transfer of small regulating RNA molecules to the neighbouring cells for maintaining cellular homeostasis.

Chapter 9
Interaction Partners of Connexin Proteins

For the functioning of proteins, they need be placed at a proper location in the cell. Connexins, being the membrane proteins, are to be delivered to the membrane for the assembly and establishment of gap junction communication. However, connexins are known to be multifunctional proteins, and the mere formation of gap junction communication does not explain all the potential functions influenced by the connexins. In addition to the gap junction communication, connexins are part of various protein complexes, which control various physiological processes of the cell. Thus, connexins not only help in intercellular communication but also mediate intracellular communication by interacting with many other proteins. The number and diversity in connexin-associating partners has increased steadily over the past years, implicating the importance of these interactions for the proper functioning of gap junctions. Below are discussed the various interaction partners of connexins and the role of these interactions in regulating various cell functions.

9.1 Cytoskeleton Proteins

For the transport and proper docking of connexins to the membrane, they need to interact with various cytoskeletal proteins. This interaction is also important for the turnover of connexin proteins. Various cytoskeleton interaction partners of connexins include microtubules, actin, α-spectrin, debrin, etc.

Microtubules are dynamic structural polymers consisting of tubulin proteins. Cx43, being one of the major connexins, is known to interact with microtubules. The C-terminal domain of Cx43 has been found essential for the interaction. The amino acids in the C-terminal domain, which are critical for this interaction, have been mapped from 234 to 243 with the sequence of KGVKDRVKGL. Cx43 binds directly to α- and β-tubulin and that microtubules at the cell periphery co-localize with Cx43-based gap junctions. Based on multiple alignment studies, similar kinds of domain have been mapped in the Cx41 and Cx46. The interaction of connexin proteins with

© Springer India 2014
M.U. Hussain, *Connexins: The Gap Junction Proteins*, SpringerBriefs
in Biochemistry and Molecular Biology, DOI 10.1007/978-81-322-1919-4_9

microtubules is essential in allowing direct transport of newly synthesized connexin hemichannels to the plasma membrane. Microtubules also interact with other junctional molecules such as cadherins, which allows increased specificity in targeting the connexins to the adherens junction, and this directs the delivery of newly synthesized connexins to areas of pre-existing adherens junctions.

Actin is the major cytoskeletal protein, and studies have revealed that Cx43 and actin co-localize in various cell types. Moreover, membrane cytoskeletal protein α-spectrin has been shown to co-immunoprecipitate with the gap junction. Another cytoskeletal protein, which has been found to interact with Cx43, is Drebin. Drebin is known to be involved in mediating cell polarity, and interestingly, studies have shown that Cx43 mediates cell migration. Whether the interaction of Cx43 with debrin is essential for mediating cell polarity and cell migration needs to be ascertained. In general, the interaction of connexin with the cytoskeletal proteins plays an important role in the life cycle of connexins by stabilization of gap junction proteins at the membrane and their internalization.

9.2 Junction Proteins

The gap junction proteins have been usually mapped in the region of plasma membrane where other junctional proteins exist. Thus, there exists a close association between the connexins and other junctional proteins. Both direct and indirect interactions of connexins with other junctional proteins have been demonstrated. The interaction between connexins and other junctional proteins is regarded to be essential for connexin assembly, trafficking, turnover, and channel gating. Hence, these interaction partners of connexin are critical for regulating the connexin life cycle, including membrane insertion, localization, and gap junctional plaque formation. At the plasma membrane, various junction proteins are known to interact with connexins, and these include Zonula occludens-1 (ZO-1), zonula occluden-2 (ZO-2), Occludin, Cadherins, and Catenins.

Zonula occludens-1 (ZO-1) and zonula occluden-2 are tight junction membrane-associated guanylate kinase proteins and are involved in the organization and trafficking of gap junctions. ZO-1 and ZO-2 interacts with the C-terminal domain of Cx43, and this interaction is mediated by the second PDZ domain of ZO-1. Similarly, studies have indicated that Cx45 interacts with ZO-1, and this interaction is mediated in a similar manner as that of Cx43. ZO-1 being a scaffolding protein can bind many other proteins, thus, increasing the potential of interaction partners of Cx43 with other proteins. Interestingly, the association between Cx43 and ZO-1 increases during certain conditions. For example, cardiac Cx43 gap junction remodelling occurs during cardiac development and in certain pathophysiological conditions. Moreover, the preferential interaction of Cx43 either with ZO-1 or with ZO-2 is determined by the stage of cell cycle. Connexin 43 interacts preferentially with ZO-1 during G0 stage, whereas the interaction with ZO-2 is preferred during the other stages of cell cycle. Based on the homology studies, the PDZ domain has been found to be present in other connexins, thus raising the potential of other connexins to interact with ZO-1/2.

Occludin is a transmembrane protein and has been found to co-localize with gap junction plaques. Freeze-fracture and co-immunoprecipitation studies have concluded that the Cx32 is a potential interaction partner of occludin. Similarly, Cx26 has been reported to interact with occludin. Interaction of connexins with occludin is regarded essential for the proper assembly of connexin at the membrane.

Adherens are another class of junctional proteins, which form close associations with the gap junction proteins. Many functions of gap junctions, like cell proliferation and cell growth, are thought to be mediated by forming functional associations with adherens. Moreover, adherens are regarded to modulate the assembly of connexins at the cell membrane. Thus, based on these functional links, it is presumed that adherens form important interaction partners of connexins. One such important interaction partner of Cx43 is β-catenin. β-catenin is a multifunctional protein, having both membrane and nuclear localization properties. β-catenin has both cell adhesion characteristics and transcriptional properties. Cx43 interacts with β-catenin and co-localizes mainly at the cell membrane. Both β-catenin and Cx43 play a role in cell growth, and the interaction of β-catenin with Cx43 is regarded as one of the mechanisms used by Cx43 in regulating cell growth. Besides Cx43, other connexins, like Cx26 and Cx32, have been shown to interact with the α-catenin. Similarly, it is established that the Cx30, Cx32, and Cx43 associate with both α-catenin and β-catenin in mammary epithelial cells. Studies have shown that this interaction is important for sequestering β-catenin away from the nucleus in differentiation permissive conditions. Besides the role of connexin–catenin interaction in regulating cell growth and differentiation, studies suggest that the interaction of α-catenin with connexins is critical for the assembly and trafficking of the gap junction proteins to the membrane.

Additionally, co-immunoprecipitation and co-localization studies have shown that N-cadherins are another type of adherens proteins that interact with Cx43. This association is regarded to be critical for the gating of Cx43 channel, as mice lacking N-cadherins show decreased cell coupling.

9.3 Enzymes

As mentioned previously, connexins are the target of various post-translational modifications, and among these, phosphorylation is the most prominent one. Most of the connexins are phosphoproteins and thus are the target of various kinases and phosphatases. The dynamics of phosphorylation and dephosphorylation regulates various aspects of connexin functions, like gating, interaction with other proteins, assembly and degradation, cell growth and differentiation, etc. For carrying out phosphorylation and dephosphorylation, it is imperative that connexins interact with various kinases and phosphatases. Various kinases are known to target connexins in general and alter their function. These kinases include tyrosine kinases and serine/threonine kinases.

The tyrosine kinases are the enzymes that phosphorylated the tyrosine residues in the target proteins. Various signalling molecules activate these kinases, and the majority of these are the growth factor signalling molecules. The binding of the signalling molecules to the receptor tyrosine kinases (RTKs) activates their intrinsic tyrosine kinase activity, which in turn results in the phosphorylation of the tyrosine residues present on the cytosolic face of these receptors. This trans-phosphorylation event of the receptor subunits and/or the associated proteins is known as trans-autophosphorylation. The intracellular signalling proteins that bind to phosphotyrosine residues on activated receptor tyrosine usually share two highly conserved non-catalytic domains called SH2 and SH3 (SH stands for Src homology regions 2 and 3 because they were first found in the Src protein, which has a role in cancer). It has been demonstrated that Cx43 is the target of various tyrosine kinases and one such known kinase is v-Src. In the *Xenopus* oocytes, mutation of the putative v-Src phosphorylation site in the Cx43 results in lack of gap junction closure by v-Src. It has been shown that the phosphorylated Tyr265 in Cx43 forms a docking site for the SH2 domain of v-Src and the SH3 domain of v-Src can bind to a proline-rich stretch in Cx43. In addition, Tyr247 can be phosphorylated by v-Src. The phosphorylation of Cx43 Tyr247 leads to the channel closure. Moreover, phosphorylated Cx43 binds to c-Src via the SH2 domain, competing with ZO-1 binding to the connexin protein.

The involvement of serine/threonine kinases in gap junction regulation has also been well documented. For example, PKC acts on Ser368 of Cx43, resulting in inhibition of gap junctional intercellular communication (GJIC). Interestingly, PKCε, which is implicated in inhibition of GJIC downstream of the fibroblast growth factor-2, is also known to co-immunoprecipitate and co-localize with Cx43 in cardiomyocytes. However, an upregulation of GJIC occurs by PKCα concomitant with Cx43 phosphorylation, and as such, maximal Cx43 phosphorylation and GJIC are observed during mammary gland differentiation and lactation. Moreover, Cx26, 32, and 43 are regulated by MAP kinases during differentiation of rat mammary epithelial cells. Interestingly, MAPK interact with Cx43 and mediate its phosphorylation; however, the effect of this phosphorylation on gap junction gating and function remains controversial. The seemingly contradictory effects of PKCs on Cx43 and GJIC likely reflect the complexity of channel regulation, where it may regulate and alter the permeability of gap junctions and hemichannels.

The phosphorylation of the connexins by various kinases demands that these events need to be regulated by dephosphorylating the target connexins. This action is mediated by the specific phosphatases. To perform such kind of reactions, the phosphatases need to interact with various connexins. In fact, numbers of interaction partners of connexins are the serine/threonine and tyrosine phosphatases. The role of serine/threonine phosphatases is to limit gap junctional conductance. These effects of phosphatases have been observed in cardiac myocytes upon conditions of hypoxia or metabolic starvation and also upon stimulation of T51B rat liver epithelial cells with EGF. Similarly, phosphatases have been shown to function upon treatments of cells with known gap junction inhibitors such as 18-β-glycyrrhetinic acid and 2,3-butanedione monoxime. On the other hand, tyrosine phosphatases have a role in enhancing GJIC, as inhibition of the activity of these enzymes significantly decreased GJIC in various cell types.

9.4   Other Proteins

Besides the associating proteins discussed so far, other interacting connexin partners are known to interact with connexins either directly or indirectly. These include plasma membrane ion channels, membrane transport proteins, receptors, aquaporin-0, and acetylcholine receptors. Various connexins have also been shown to interact with calmodulin, which has been implicated in control of connexin channel gating. Furthermore, the calcium-/calmodulin-dependant kinase II (CaMKII) interacts with and phosphorylates Cx36 and mediates channel gating. CaMKII has also been shown to mediate enhanced gap junctional coupling, mostly through Cx43, in response to high extracellular K^+ in mouse spinal cord astrocytes. Thus, this interaction of connexins with CaMKII may have a general regulatory role in neuronal signal transmission, with a role in electrical coupling in addition to the defined role of CaMKII in chemical synaptic transmission. Other interactions include cholesterol, which was shown to interact with gap junction channels in different ratios during fibre cell differentiation in embryonic chicken lens. Moreover, Cx43 is known to translocate to mitochondria and interacts there with the heat shock protein 90 (Hsp90) and translocase of the outer mitochondrial membrane (TOM).

Furthermore, we have recently shown that Cx43 interacts with an adenosine triphosphate-sensitive K^+ channel (Kir6.1). Adenosine triphosphate-sensitive K^+ channels are the protein channels responsive to changes in ATP/ADP ratio and provide a means to couple movement of potassium ions, to relate membrane potential with cellular energy status. Both Cx43 and Kir6.1 are known for their role in cell protection during stress conditions. Moreover, studies have shown that Cx43 and Kir6.1 are part of the same signalling pathway that imparts protection during ischaemia preconditioning. Interestingly, it was shown that phosphorylation at Serine 262 of Cx43 was indispensible for the interaction with Kir6.1. Studies have shown that the status of Cx43 phosphorylation is regulated in response to various physiological functions and has been attributed to alter interaction of Cx43 with other proteins. Moreover, phosphorylation of Cx43 has been shown to be indispensable for various protective treatments such as ischaemic preconditioning.

Chapter 10
Tissue Distribution of Connexins

There are more than 20 known connexin proteins identified in the human and mice genome. However, not all the 20 connexins are expressed in all the cells. The expression of connexins has been shown to vary between tissues. Some of the connexins, for example, Cx43, show wide tissue distribution and while others are restricted to a particular cell type or tissue. The tissue distribution of connexins demonstrates their relevance in executing tissue-specific functions. Although connexins show similar, topological features, there exists appreciable amount of variability among different connexins. The variability involved in connexins allows for a great deal of diversity in gap junction formation. Each gap junction appears to confer some specificity for what type of molecules pass through it, based on either the charge or size of the molecule. Based on that specificity, it seems likely that even small amounts of a particular gap junction with a unique composition of connexins could be important for the movement of a particular metabolite or set of metabolites. Identifying what connexins are present in a particular tissue, even if only in small amounts, could thus be crucial for understanding their roles in cell communication as well as cell adhesion. These subtle variations in connexins and hence gap junctions are crucial for performing specialized function of different cell types and tissues. In the following paragraphs, the tissue distribution and tissue-specific function of connexins will be discussed.

10.1 Connexins in Vascular System

The vascular system is known to expresses many connexins, like Cx37, Cx40, Cx43, and Cx45. Both the smooth muscle cells and the endothelium cells express Cx37, Cx40, and Cx43. Using immunocytochemistry, Western blotting, and electron microscope studies, the presence of Cx37 and Cx40 has been unequivocally found in the endothelium. However, the expression of Cx43 in the endothelial cells is more controversial and appears to vary with vessel size, vascular region, and species.

© Springer India 2014
M.U. Hussain, *Connexins: The Gap Junction Proteins*, SpringerBriefs
in Biochemistry and Molecular Biology, DOI 10.1007/978-81-322-1919-4_10

In the smooth muscles, connexin expression is not that well defined as that of the endothelial cells. Expression of Cx43 and Cx40 in the smooth muscle cells has been described, and recently, Cx45 has been reported as well. Moreover, in some instances, Cx37 that is usually thought to be an endothelial connexin has also been reported in smooth muscle cells. Physiological significance of connexins in the vasculature is to coordinate various functions by communicating between different cells. This communication between different cells of vasculature plays a central role in coordinating various cellular functions. Moreover, gap junctions formed between the smooth muscle and endothelium provide the pathway for the radial and longitudinal communication in the vascular system. Physiological significance of various vasculature connexins is briefly discussed below.

Cx40: The role of Cx40 has been demonstrated using Cx40 knock-out animals. Based on these studies, it has been shown that Cx40 is involved in the vasomotor tone regulation. Cx40-deficient animals develop hypertension and depict irregular vasomotion. The occurrence of hypertension in these animals has been correlated with the reduced density of the endothelial gap junctions, mediated by Cx40. The role of Cx40 in regulating the blood pressure has been attributed due to the formation of gap junctions between the renal endothelial cells and the renin-secreting cells of the afferent arteriole. Thus, the intercellular communication between the endothelial cells and the renal renin–angiotensin system plays an important role in the regulation of blood pressure. Moreover, deletion of Cx40 results in the elimination of a very rapid, non-decremented component of axial conduction induced by electrical stimulation or acetylcholine (Ach) stimulation. More studies are required to understand the molecular basis for this type of conduction. This will pave way for the development of various vasodilatation methods, enabling to communicate over long distances using Cx40 gap junction channels and thus might play a key role in the maintenance of vasomotor tone and blood pressure.

Cx45: The role of Cx45 in the vasculature has been ascertained using Cx45-deficient animals. Although the Cx45-deficient embryos show normal initiation of vasculogenesis, however, various defects are manifested, which include defective remodelling and organization of blood vessels and failure to form a smooth muscle layer surrounding the major arteries.

Cx37: Although Cx37 has been shown to be expressed in the vascular smooth muscles, deletion of these connexins does not show any defective vasculature or any defect related to the blood vessel development. Interestingly, simultaneous deletion of Cx37 and Cx40 is lethal with acute vascular abnormalities.

Cx43: Although Cx43 expression in vasculature is not that intense, it has been found to be important for the proper development of the vascular system. Cx43 controls cell proliferation and migration, and its expression in the smooth muscle cells is induced during mechanical injury. The role of Cx43 in cell migration, after mechanical injury, has been corroborated in the cultured endothelial cells, which resulted in the increased expression of the Cx43 expression, with the downregulation of the Cx37 expression, and no change in the Cx40. Moreover, endothelial cell-specific deletion of Cx43 causes hypotension, in contrast to the deletion of Cx40, which results in hypertension.

10.2   Connexins in the Heart

The heart is an electric organ and the expression of gap junctions in this organ is well suited for the physiological function of the heart. For the conduction of electric impulses in the heart, the intercellular gap junction channels between various cells are regarded indispensible. The presence of different connexins in the heart is crucial for the functional differences between various regions of the heart. Although the different connexins expressed in the cardiomyocytes form similar gap junction channels, these gap junctions also differ significantly in their channel properties in accordance to the region where they express.

The gap junction channel properties in the cardiomyocytes are highly dynamic and are targets of modulation by several conditions, for example, hypoxia/ischaemia. The short-term changes on the gap junctional conductance are mainly caused by phosphorylation. Moreover, the prolonged changes in the gap junctions are known to occur at the transcriptional level. Various growth-inducing factors, like epidermal growth factor (EGF), platelet-derived growth factor (PDGF), transforming growth factor b (TGFb), and tumour necrosis factor a (TNFa), are reportedly known to increase the expression of Cx43 in the neonatal cardiomyocytes. The increased expression has been attributed to the increase in the mRNA levels as well as in the decrease in the turnover of Cx43.

The expression of the connexins varies between different regions of the heart and during the development of the heart. Connexin43 (Cx43) constitute the major gap junction protein that is expressed in the cardiomyocytes of atrial and ventricular mammalian myocardium. In the mouse and rat, it is abundantly expressed in all the cardiac compartments with the exception of the sino-atrial (SA) and atrioventricular (AV) nodes, the His bundle, and proximal parts of both bundle branches, but it is expressed in more distal parts of the bundle branches. In humans, Cx43 is expressed in all parts of the ventricular conduction system. The role of the Cx43 in the normal functioning of the heart can be ascertained by the fact that many genetic alterations of the Cx43 in mouse lead to various abnormalities. For example, disruption of both the alleles of Cx43 in mouse embryos results in their death shortly after birth. The cause of death is due to the asphyxiation caused by the obstruction of right ventricular outflow. Moreover, Cx43 knock-out mice models have demonstrated the role of Cx43 during the development of the heart. The Cx43 null mice embryos show delayed looping of the heart tube and the formation of certain bulges, lined by smooth muscle marker cells. The abnormal bulges have been demonstrated to be formed due to the failure of cardiac crest cells to migrate to the tubular heart to form proper epicardium. The migratory failure of these cells has been attributed due to the lack of Cx43 gap junction communication. These studies have been corroborated by the transgenic mice that over-express Cx43 and show increased migration of cardiac crest cells, while the mice expressing low levels of Cx43 showed decreased migration of these cells. In addition, Cx43 null mice show abnormal development of coronary arteries and have reduced diameter, decreased expression of myosin, etc. A similar kind of manifestation has been observed in various human cardiac abnormalities, including the coronary artery anomalies. Interestingly, various genetic mutations in human Cx43 have been associated with various coronary artery abnormalities, and the deficiency of Cx43 in the ventricular cardiomyocytes has been associated with arrhythmias and sudden death. One of the common conditions associated with the embryos of Cx43 null mice−/− is the arrhythmias, and when these hearts are exposed to acute ischaemia, they show high frequency of ventricular tachyarrhythmias as opposed to heterozygous Cx43+/− mice. Thus, reduction of Cx43 expression, and consequently the electrical coupling, may play a critical role in ventricular arrhythmogenesis. Remodelling of Cx43, during acute ischaemia, is known to be the primary cause of developing arrhythmogenesis.

The Cx40 is expressed in the fetal and neonatal ventricles; however, its expression goes down drastically during the adult stages. Low levels of Cx40 expression have also been found in the SA node and AV node. Cx40 is involved in the propagation of electrical impulse from the atria to the ventricles. The role of Cx40 has been ascertained using knock-out mice. Mice heterozygous for Cx40 deletion possess similar electrophysiological properties as compared to wild-type mice. However, Cx40 KO null mice (Cx40−/−) show many electrophysiological disturbances. For example, the Cx40 KO mice exhibit disturbed cardiac influx propagation at various levels of the CCS and an increased incidence of inducible atrial arrhythmias. The disturbances of impulse conduction are in agreement with the expression localization of Cx40 in the AV node, the His bundle, and the bundle branches. In addition to the alterations of the electric activity, Cx40 KO mice show other cardiac malformations. These defects include atrial and ventricular septation and in some cases hypertrophy. Moreover, deeper analysis of Cx40 heterozygous newborn mice showed bifid atrial appendages, ventricular septal defects, tetralogy of Fallot, and aortic arch abnormalities, whereas the Cx40 KO (Cx40−/−) mice showed double-outlet right ventricle, tetralogy of Fallot, and partial endocardial cushion defects. The defects in Cx40−/− indicate that Cx40 has a role in septum formation and in other cardiac development events. The expression pattern of Cx40 in the human heart is similar to that of the mouse heart. Thus, it is expected that the alterations in the Cx40 expression in humans will have similar consequences as that of Cx40 KO mice. Interestingly, dominant mutations in the transcription factors Tbx5 and Nkx2.5 that regulate the expression of Cx40 result in alterations in the cardiac electrophysiological and morphological phenotype resembling that of the Cx40-deficient mice.

As far as Cx45 is concerned, its expression has been observed in the SA node, AV node, and the ventricular conduction system. Moreover, low levels of expression have been also reported from the atria and the ventricles. The importance of Cx45 can be gauged by the observation that the Cx45 KO mice (Cx45−/−) die in utero on day 10.5 pc. The death is known to be caused by the defective vascular development and the block in the atrioventricular conduction. The Cx45 is regarded essential for the embryonic heart development. During the early stage of development, Cx45 expression is seen in most of the cardiac compartments, and in the later stages, its expression goes down and remains specific to particular regions. Hence, in adult heart, Cx45 expression remains confined to SA node and inter-ventricular septum.

10.3 Connexins in the Nervous System

Connexins were first described as the important molecular components of the electrical synapses in the central nervous system (CNS). Connexin expression is widely distributed in the central nervous system. Besides forming electrical synapses, connexins are also important as far as other physiological functions of the brain are concerned. In the following paragraphs, expression and functional analysis of different connexins in different cell types of the central nervous system will be discussed.

10.3.1 Neuronal Connexins

Different techniques were instrumental in ascertaining the number of connexins, which are expressed in different regions of the central nervous system (CNS). Of these, electron microscopy and immunocytochemistry were very helpful in providing insights about the expression of different connexins in diverse regions of the CNS. Furthermore, with the use of different antibodies raised against various connexin proteins, it was possible to immuno-detect various connexins in the different regions of the CNS. Until now, more than eight different connexins has been detected in different neuronal cell populations. Cx32 is one of the major neuronal connexins expressed in the neurons. Moreover, Cx26, Cx36, and Cx43 have been detected in different neuronal subpopulation. Connexins have been shown to be widely expressed in different regions of the brain. These regions include the hypothalamus, striatum, inferior olive, hippocampus, olfactory bulb, retina, cerebral cortex, and cerebellum. Generally, it is regarded that the mature brain expresses less connexins as compared to the developing brain and there is a progressive decrease in connexin expression and gap junctional intercellular communication (GJIC) with the maturation of brain, although neuronal coupling persists in many brain regions in adults. Thus, it is regarded that there exists an inverse correlation between the connexin expression and neuronal differentiation. The observation is suggestive that gap junctional communication between different cells of the central nervous system allows passage of different signalling molecules between neurons and thus regulates the neuronal differentiation. Besides specific expression of connexin in different regions of the brain, there also exists neuronal specificity of connexin expression. Connexin expression and gap junction communication has been found to exist in different neuronal cell types, like cerebellar basket cells, pyramidal cells in cerebral cortex and hippocampus, medium spiny neurons in the striatum, dopaminergic neurons in the substantia nigra, and motor neurons in the spinal cord. Moreover, the GABAergic interneurons are interconnected through electrical synapses, and this includes interneurons from the cerebral cortex, cerebellum, and striatum and reticular neurons from the thalamus. Pyramidal cells of CA1 and CA3 regions of the hippocampus communicate with each other using ultrafast axo-axonal coupling, and this is achieved using gap junction communication mediated by connexins.

10.3.2   Glial Connexins

In the CNS, glial cells outnumber the neuronal cells, and most of the glial cells are coupled through the gap junctions. Connexin expression remains more or less consistent in the glial cells during the development stages and persists through-out the differentiated stage. Connexin expression in various glial cell types is discussed below.

10.3.2.1   Astrocytes

Astrocytes are the star-shaped glial cells with long processes. The syncytial organization of astrocytes is maintained through the gap junction communication. The interconnection of the astrocytic organization can be appreciated by injecting a low molecular weight tracer, like lucifer yellow, into a single astrocyte, and within no time, the dye appears in around 100 astrocytes. Using various gap junction-specific inhibitors, like carbenoxolone, the dye spread remains restricted to the injected astrocytes, thus confirming the role of the gap junction communication in maintaining the astrocytic organization. The presence of high abundance of gap junctions in the astrocytes allows direct intercellular diffusion of ions, nutrients, and signalling molecules between these cells. Cx43 is the most abundant connexin expressed in the astrocytes and thus constitutes the major connexin contributing to the gap junctional communication of the astrocytes. Moreover, low levels of Cx30, Cx40, and Cx45 have been also detected in the astrocytes. Cx43 expression in astrocytes starts very early during development, and the levels increase progressively during the adult stage.

10.3.2.2   Oligodendrocytes

Oligodendrocytes constitute the myelin-forming cells of the central nervous system and are known to possess gap junctions. One of the major connexins expressed in the oligodendrocytes is Cx32. Moreover, Cx45 has also been detected in the oligodendrocytes. Immunofluorescence studies have shown that the Cx45 is expressed mainly in the cell soma and proximal processes of oligodendrocytes localized in the white and grey matter. Recently, a novel gap junction protein called Cx29 was detected in the oligodendrocytes. The expression of different connexins in the oligodendrocytes is indicative of the presence of heterologous gap junctions between the oligodendrocytes. In fact, the presence of homologous gap junctions between the oligodendrocytes is still debated, and accumulating evidence indicates that the majority of gap junctions in oligodendrocytes are heterologous.

10.3.2.3 Microglial Cells

Microglia belongs to the class of glial cells that provide macrophagic function in the brain and spinal cord. The microglia constitutes the first line of immune defence in the central nervous system (CNS). The microglia constitutes 10–15 % of the total glial cell population within the brain. Under normal conditions, the microglia acquire a specific ramified morphological phenotype termed "resting microglia". Microglia provides scavenging function in the CNS against damaged neurons and infectious agents. The microglia perform an important function in the CNS by acting as antigenic presenting cells and thus activating various immune cells. Microglial cells are considered the most susceptible sensors of brain pathology. Upon any detection of signs for brain lesions or nervous system dysfunction, microglial cells undergo a complex, multistage activation process that converts them into the "activated microglial cell". For the activation, the microglia utilizes various communication mechanisms, and one such signalling communication is provided by the gap junction channels. Thus, it is regarded that the microglia uses gap junctional communication as an important means to achieve the activation state induced by specific factors in their microenvironment. Connexin expression in the microglial cells has been well documented. The expression of connexins in the microglial cells is highly dynamic. In the normal adult rat cerebral cortex, less than 5 % of microglial cells are found to be Cx43 immuno-reactive. Moreover, the in vitro cultures of microglial cells show low levels of diffused Cx43 levels. However, in both cases, Cx43 expression dramatically increases after the activation of microglial cells. The activation under in vivo conditions can be caused by the stab wound, while in culture after treatment with inflammatory cytokines. Moreover, under culture conditions, treatment with cytokines also shows increased cytoplasmic to membrane localization of the Cx43 and thus increased connectivity between the activated microglial cells by gap junctions. Besides Cx43, it is regarded that other connexins mediate gap junctional communication between the microglial cells. These observations are based on the fact that the microglia of Cx43 null mice retain the capability of gap junctional communication.

10.3.2.4 Ependymal Cells

Ependymal cells are the specialized glial cells that form the lining of the ventricles and cerebral aqueducts. These cells are highly coupled with the gap junctions. Ultrastructural details of the ependymal cells show numerous gap junctions, with smaller gap junctional plaques localized at the apical margins and the larger plaques at the lateral membranes between the apposed cells. The efficient coupling is important for the synchronized activity of ciliated ependymal cells. Connexin proteins that are specific to the ependymal cells are Cx26 and Cx43.

10.3.2.5   Meningeal Cells

The CNS is surrounded by three protective connective tissue sheaths of mesenchymal origin. Dura mater is the external meninge, whereas the two inner ones are the arachnoid and pia mater, or the leptomeninges, which send extensions into the neural parenchyma. Gap junctions are extremely abundant in the developing and adult meninges. Indeed, cultured leptomeninges are strongly coupled, even more than in astrocytes. Three connexin types, Cx26, Cx30, and Cx43, are expressed at high levels in meningeal cells which show strong punctate staining.

10.3.3   Gap Junctions Between the Glial Cell Types

Besides forming gap junctions between its own cell types, numerous gap junctions are also known to exist between different glial cell types. For example, electron microscopy studies have indicated that there exist numerous gap junctions between astrocytes and oligodendrocytes. In addition, functional studies have demonstrated the occurrence of gap junction communication in the cocultures of astrocytes and oligodendrocytes. Gap junctions between these two glial cell types occur at specialized regions along the surface of these cells. The major connections occur between the two cell bodies, between soma and glial processes, and between astrocytic processes and the external foil of the myelin sheath of oligodendrocytes. The main connexins that contribute to these gap junctions are the Cx43 and Cx32.

10.3.4   Gap Junctions Between the Glial Cells and Neurons

The existence of gap junction communication between glial cells and neurons is conditional rather than absolute. This type of heterocellular interactions occurs only under certain conditions. For example, in the cocultures of neurons and astrocytes, gap junction plaques are detected for a certain time window. During this time, it is regarded that the gap junctional communication is important for the neuronal differentiation. Although the connexins involved in this communication are not fully known, locus ceruleus glial cells express Cx26, Cx32, and Cx43, whereas neurons express Cx26 and Cx32. Accordingly, the formation of homomeric and/or heteromeric channels could be involved in heterocellular coupling.

The gap junctional permeability in neurons and astrocytes is highly dynamic. For the functioning of gap junctions in the brain in a dynamic manner, they are regulated by a number of bioactive molecules. Similar to other tissues, gap junctions in the CNS are subjected to long- and short-term regulation. Long-term regulation occurs over hours or days and operates at the transcriptional level. This is associated with changes in the expression of connexins and thus the number of junctional plaques. Short-term regulation occurs in minutes and deals with changes in opening probability, time of opening/closure, and/or unitary conductance of functional channels already in place at gap junctions. Moreover, changes in the rate of gap junction internalization and connexin degradation may also occur. In addition, connexin expression and hence gap junctional communication in the astrocytes and in neurons is regulated by neurotransmitters, growth factors, peptides, cytokines, and endogenous bioactive lipids. Thus, like chemical synapses, electrical synapses, mediated by gap junctions in the brain, are subjected to some plasticity and are tightly modulated by neuronal products and secretions from other brain cell types, such as glia (astrocytes, microglia) and endothelial cells. Hence, it is suggested that any abnormal release of these compounds or signals results in the change of connexin expression. These changes are often responsible for various pathophysiological disorders of the brain. Numerous bioactive compounds, like monoamines, excitatory and inhibitory amino acids, and their derivatives, are known to regulate the gap junctional communication. The effect of bioactive amines is mediated through the generation of various secondary messengers and protein kinases, and these actions can be mimicked by direct activation of several transduction signalling pathways using various agonists. The use of various agonists and antagonists and their effect on the gap junctional communication has paved way for using various gap junction-specific drugs. In fact, alteration of gap junction communication between neurons has been observed in vivo after applying antipsychotic drugs or amphetamine withdrawal, suggesting that electrical synapses and neuronal connexin are considered as potential targets for drug therapies. Moreover, like chemical synapsis, synaptic efficacy at electrical synapses is potentiated using various signalling inputs. For example, use of electrotonic transmission at excitatory inputs to the goldfish Mauthner cell exhibits a long-term potentiation due to an increase in gap junctional conductance produced by a stimulation paradigm similar to that used in the hippocampus.

10.3.5   Functions of Neuronal Connexins

10.3.5.1   Electrical Transmission

The gap junction formed by the neuronal connexins is called the electric synapses. Most of the electrical synapses analyzed are characterized by their bi-directionality, voltage dependency, and a low average coupling coefficient. Electrical synapses are efficient in transmitting pre-potential spikes and after hyperpolarization phases, as compared to a single action potential. Electrical synapses act as low-pass filters, and their cut-off frequency explains the functional involvement at the neuronal network level. There are instances in the CNS where electrical and chemical synapses coexist and modulate each other's function. For example, coexistence of electrical and chemical synapses in inhibitory neuronal networks allows for enhanced timing of spike transmission (1–2 ms for mixed synapses versus 10–20 ms for electrical coupling alone). Electrical synapses are known to contribute to several functional network properties of the brain, like spike synchronization and network oscillations, coordination and reinforcement of postsynaptic inhibitory potentials, and the detection of coincidence in the inhibitory networks. Electric synapses play a role in the gamma oscillation of the brain, and these are mediated by the Cx36. Targeted deletions of Cx36 show reduced synchrony of gamma oscillations (30–70 Hz) in the neocortical slices. However, ultrafast oscillations or "ripples" (~150 Hz) remained unaffected in such knock-out animals. Other connexins are known to be part of the electric synapses and play role in various physiological functions of brain. Besides their electrical role at neuronal synapses, gap junctions are also important for the metabolic and biochemical coupling by allowing the passage of various molecules. This contribution of gap junctions to non-synaptic interactions between neurons involves the passage of low molecular weight signalling molecules, such as secondary messengers (cyclic nucleotides), amino acids (glutamate), and metabolites (glucose, lactate). Passage of signalling molecules is regarded important for the proper maturation of the developing brain. For example, the passage of inositol trisphosphate (IP3) through the gap junction of developing visual cortex neurons is important for coordinating their neuronal activity.

10.3.5.2   Homeostasis

Besides electrical transmission, gap junctions are critical for the neuronal homeostasis. During neuronal activity, increased concentration of potassium [K^+] in the external environment of neurons can induce shrinkage of the glial cells. Glial cells maintain the normal neuronal activity by rapid withdrawal of the increased potassium concentration from the extracellular environment. Glial cells achieve this potassium buffering by having gap junctions between themselves. Thus, gap junctions constitute an intercellular pathway that transfers K^+ from areas of high concentration to those having lower concentration. Gap junctions in the astrocytes are also involved in the homogenization of intracellular ionic concentrations. Inhibition of gap junction in cultured astrocytes results in the imbalance of intracellular Na^+ concentration.

10.3.5.3   Energy Supply to Neurons

Gap junctions form channels through which various metabolites can easily pass. Glucose, glucose-6-phosphate, and lactate are well known to pass through the astrocytic gap junction channels. Astrocytes contain long processes that link the endothelial cells of blood capillaries at one end with the neuron at the other end. The astrocytes form important intracellular pathways for the transfer of energy metabolites from blood to neurons. Gap junctions contribute to these pathways by providing neurons with energy-producing compounds, since astrocytes are essential morphological intermediates located between blood capillaries and neurons. Moreover, during certain hypoxic conditions, high concentration of metabolites in the astrocytes provides important reservoirs to provide glucose and lactate to the neurons via gap junctions.

10.3.5.4 Neuroprotection

The role of gap junctions in protecting neuronal cells from ischaemic insults is well established. The neuroprotective role of the connexins has been established by the studies showing that the closure of gap junction channels by uncoupling agents results in the increased neuronal vulnerability to ischaemic insults. Moreover, in agreement with this hypothesis, Cx43 heterozygote knock-out mice, ischaemia induced by the occlusion of the middle cerebral artery shows a large infarct volume as compared to that observed in the wild-type mice. These studies confirm the role of gap junctional communication in the neuroprotection. However, there are also some reports that the gap junction channels in the astrocytic network propagate death signals from the site of injury insults to other parts of the CNS.

10.3.5.5 Regulation of Cell Volume

Regulation of cell volume is of primary importance for the normal functioning of the brain. Brain cells are endowed with the capability to regulate cell volume during certain pathophysiological conditions. Exposure of astrocytes to hyposmotic solution results in transient changes in their cell volume. However, immediate response of the astrocytes to the hyposmotic conditions is associated with an increased conductance of the gap junctions. Moreover, astrocytes restore their volume by losing ions and amino acids, and osmotically regulated water and gap junctional communication are important for such functions.

10.3.5.6 Propagation of Intercellular Calcium Waves

It is well established that the astrocytes propagate intercellular Ca^{2+} waves over long distances in response to stimulation and, similar to neurons, release transmitters, called "gliotransmitters", in a Ca^{2+}-dependent manner. Calcium signals and the occurrence of calcium waves in astrocytes provide these cells with a specific form of excitability. Various lines of evidence have shown that there exist different pathways for the transmission of Ca^{2+} waves in the astrocytes. Some of these involve the direct communication between the cytosols of two adjoining cells through gap junction channels, while others depend upon the release of gliotransmitters that activate membrane receptors on neighbouring cells. Gap junction-mediated transmission of Ca^{2+} waves was first identified in astrocytes. The conclusion was drawn from the use of gap junction communication inhibitors that impaired the Ca^{2+} wave spread across the astrocytic network. This finding together with several other studies provided a strong basis in support of the view that the gap junction channels play a crucial role in the transmission of Ca^{2+} signals between astrocytes. Besides playing a role in Ca^{2+} wave propagation between the astrocytes, it is also established that the astrocytes signal to neurons through Ca^{2+}-dependent release of glutamate. Various studies have indicated that the Cx43 and Cx32 have the ability to pass

intercellular Ca^{2+} waves between the astrocytes and the neurons. Interestingly, studies have pointed the role of IP_3 that can pass via gap junction channels and thus initiate the release of Ca^{2+} in the neighbouring cells. Thus, it is regarded that IP_3 per se and not Ca^{2+} results in the direct propagation of Ca^{2+} waves. Besides direct gap junctional communication, connexin hemichannel-mediated Ca^{2+} wave initiation has also been proposed. However, the involvement of connexin hemichannels in Ca^{2+} wave propagation is regarded to be indirect rather than a direct effect. It is proposed that the purinergic receptor-mediated ATP release is controlled by connexins and the release of ATP initiates Ca^{2+} wave.

10.3.6   Connexin Remodelling During Brain Pathologies

Various central nervous system-associated injuries and pathologies are associated with the modulation of gap junctional communication and connexin expression. The regulation of connexin expression during brain pathologies is a cause or consequence of such conditions and remains to be elucidated. In the following paragraphs, modulation of connexin expression in various brain pathological situations will be discussed.

10.3.6.1   Brain Inflammation

Brain inflammation occurs under various conditions, which include traumatic injury and brain diseases such as multiple sclerosis, ischaemia, and Alzheimer's disease. One of the hallmarks of brain inflammation is a condition called reactive gliosis. Reactive gliosis results in the proliferation of glial cells (astrocytes and microglia) to lesion site and is characterized by the glial fibrillary acidic protein (GFAP)-positive astrocytes. In addition, these conditions are accompanied by the modulation of gap junction communication. Reactive gliosis has been associated with modulation of Cx43 expression in the astrocytes and hence the gap junction communication. Moreover, during certain brain diseases, like Alzheimer's, an increased Cx43 immunoreactivity is observed at sites containing amyloid plaques. Moreover, during inflammation, activation of endothelin receptors by the endothelin results in the inhibition of gap junctional communication and hence propagation of calcium waves. Interestingly, during reactive gliosis, most of the components of the endothelin system (endothelin, endothelin receptors, endothelin-converting enzymes) are upregulated. In addition, an endothelin level is increased in several neurological disorders, such as Alzheimer's disease, subarachnoid haemorrhage, and ischaemia. The astrocytic gap junctional communication is also modulated by the release of nitric oxide, produced by inducible nitric oxide synthase and prostaglandin (PGE2), produced by the cyclo-oxygenase activation. All of these compounds contribute to the glia-mediated neuro-inflammatory response by affecting gap junction communication. The conclusion drawn from these observations suggests that during local inflammation in

the brain, pro-inflammatory cytokines, endothelins, etc., regulate the gap junctional communication of the astrocytes. This has been corroborated by the downregulation of Cx43 in the astrocytes under inflammatory conditions. Although the functional consequences of gap junction inhibition in the astrocytes during inflammation are not understood fully, it is regarded that inhibition of gap junctional communication may restrict the passage of active molecules to neighbouring astrocytes. This will reduce the spread of apoptotic signals within astrocytic networks and thus isolate intact tissues from primary lesion sites. However, it is pertinent to mention here that under certain cell-damaging conditions, increase in the gap junctional communication is known to dilute the damaging signal and thus contribute to the bystander effect and hence decrease neuronal vulnerability to oxidative stress. Therefore, reactive astrocytes with modified gap junctional communication are regarded as a key response in a dynamically changing environment that can modify neuronal functions and overall brain physiology.

10.3.6.2 Epilepsy

Epilepsy is characterized by recurrent seizures and results from abnormal synchronous firing of neurons. Typically, it originates in networks that under normal conditions generate local or large-scale synchronized oscillations. Multiple factors contribute to this activity, including strong recurrent excitatory connections, the presence of intrinsically burst generated neurons, and ion regulation. Gap junctions or electric synapses play a critical role in the generation and propagation of various oscillatory waves in the central nervous system. Simulation and modelling of neuronal networks have supported the importance of GJIC in synchronized activity and how electrical coupling can modify frequency of oscillations and firing properties of neurons. Thus, gap junction-mediated electric synapses are crucial for synchrony and stabilization of bursting firing patterns of the neurons. The role of connexins or gap junctional communication has been demonstrated using various animal model studies. For example, Cx36-null mice show deficient synchronous activity of inhibitory interneuronal networks in neocortex. Moreover, these mice show impaired hippocampal gamma (~30–80 Hz) oscillations. Similarly, Cx32 knock-out mice show myelination defects with neuronal hyper-excitability. Recently, it has been demonstrated that very fast neuronal oscillations (VFOs, 140–200 Hz) are involved in the generation of seizures. These very fast neuronal oscillations have been shown to immediately precede seizure onset. Interestingly, studies have indicated that the axo-axonal gap junctions are involved in the generation and propagation of high frequency oscillations. It has been found that a very small number of gap junctions linking neurons are required to produce the appropriate oscillatory activity. These studies were confirmed by the electrophysiological detection of the axo-axonal gap junctions between CA1 pyramidal neurons. Moreover, in rat hippocampal, use of gap junction inhibitor carbenoxolone abolished very high frequency generated before epileptiform bursts. Similarly, in vivo evidence with the cat neocortex has shown that halothane (gap junction inhibitor) prevents the onset of ripples observed during the epileptic seizures. These studies have confirmed the potential of gap junction inhibitors as pharmacological agents that may prove effective anticonvulsants.

10.3.6.3   Brain Ischaemia

Ischaemic brain injury is one of the major causes of neurological malfunctioning. Brain ischaemia can engulf the major portion of brain and hence is named as global ischaemia, or it can be localized, called as focal ischaemia. Global ischaemia is usually caused when the blood supply to the brain stops temporarily, due to either cardiac arrest or systemic circulatory collapse. The resulting insults are of short duration with little or no blood flow changes. In contrast, focal ischaemic injury usually occurs if the blood flow to a particular region of the brain decreases, thus affecting its functioning. Few of the major outcomes of the focal ischaemia is that the cellular energy is depleted within minutes, there is a sudden loss of specific brain functions, and a core of dying tissue, the "infarct", develops. Besides neuronal damage, the ischaemic insult results in the swelling of astrocytes and malfunctioning of the glutamate uptake. Besides, the connexin expression in the astrocytes is modulated, and this results in the disturbed gap junctional communication. Since gap junction communication is pivotal for the normal functioning of neurons and brain homeostasis, hence, altered astrocytic gap junctions contribute to the neuronal death. Moreover, gap junction can act as a medium to propagate the secondary expansion of focal ischaemic injury. It has been observed that the astrocytic gap junctions remain open during ischaemia and mediate the propagation of cell death signals to the other parts of the brain. Modulation of gap junction communication after brain ischaemia is associated with the change in the expression of connexins. These changes can be either the upregulation of certain connexins or the downregulation of others. For example, in the hippocampus, global ischaemia induced an increase in Cx32 and Cx36 proteins, specific to the inhibitory interneurons of the CA1 region, whereas in CA3 region the expression of Cx32 and Cx36 in the neurons and the Cx43 expression in the astrocytes remain unchanged. It has been proposed that the increase of Cx32 and Cx36 expression in CA1 region contributes to the survival of the GABAergic neurons and increases their synchronized inhibitory synaptic transmission. Similarly, ischaemia of the forebrain, induced by bilateral carotid occlusion, produced increased Cx43 immunoreactivity at sites of mild injury, whereas regions exhibiting severe ischaemic injury showed a decreased Cx43 immunoreactivity.

10.4   Connexins in the Skeletal System

Skeletal systems have abundant gap junctions present in all the bone cells, with osteoblasts and osteoclasts having the highest number. The presence of abundant gap junctions in the bone cells is suggestive of their involvement in various bone functions, including control of osteoblastic cell proliferation, differentiation, and survival. Although many connexins are known to express in the skeletal system, Cx43 forms the most abundant gap junctions in the skeletal system. Other connexins, which are expressed in the skeletal system, are Cx45 and Cx46. Besides, Cx40 have been shown to be present in the developing limbs, ribs, and sternum, but its expression goes down with the maturations, and in the adult skeletal system, its expression is not documented. Cx43 being the highly expressed connexin in the bone, its biological importance in the skeletal development has been established by numerous studies using human and mouse genetics. Mutational studies in mice with germ line null mutation of Cx43 indicated the hypo-mineralization of craniofacial bones and a severe delay in ossification of the axial and appendicular skeleton. In addition, numerous other skeletal abnormalities occur due to the absence of Cx43, like ossification defects and malformation of cranial ribs, vertebrae, and limbs. Moreover, osteoblast-specific deletion of Cx43 in the mice shows similar defects, excluding the craniofacial malformations or the ossification defects. In humans, the linkage of mutation of Cx43 locus to the human disease called oculodentodigital dysplasia (ODDD) provides the strongest evidence for a critical role of Cx43 in skeletal development. The molecular mechanisms of Cx43 action on the bone metabolism are still not well understood. However, recent studies have indicated that Cx43 mediate bone cell response to the hormonal stimulation. For example, the anabolic effect of parathyroid hormone (PTH) on the bone is attenuated in Cx43-deficient mice. Further analysis has shown that Cx43 deficiency results in the diminished production of PTH stimulated cAMP and hence decreased mineralization of the osteoblasts. In addition, as a mediator of hormonal stimulus, Cx43 also plays a role in the anabolic response to the mechanical stimulus. It has been observed that the mineral deposition rate at the mechanically stimulated endocortical surface of tibiae is significantly reduced in the conditionally Cx43 deleted mice relative to wild-type animals. Besides the involvement of Cx43 in regulating various physiological aspects of bone metabolism, its role in mediating the effect of various skeletal pharmacologic agents has been proposed. For example, the inhibitory action of bisphosphonate and alendronate on the apoptosis of osteoclasts has been shown to require Cx43. It has been observed that the anti-apoptotic action of bisphosphonates is independent of Cx43 gap junction communication. However, the role of Cx43 hemichannels has been proposed to be required for such effects. Cx43 is also known to modulate the expression of various genes in the osteoblastic cells. The transcription of $\alpha_1(I)$ collagen and osteocalcin has been shown to be influenced by the expression of Cx43. Cx43 has been shown to exert its influence on the transcription of various genes by specific DNA promoter elements, known as "connexin response elements". Similarly, Cx43 mediates its effect on the transcription of osteocalcin and $\alpha 1(I)$ collagen genes through Cx43 response elements and the binding of Sp1/Sp3 transcription factors. Cx43 regulate gene expression either directly or indirectly by altering various signalling pathways. In osteoblasts, Cx43 alters ERK signalling, and this in turn modulates gene transcription from osteoblast gene promoters via decrease of ERK-dependent phosphorylation of Sp1 with preferential recruitment of Sp3 to connexin response elements. Cx43 also mediate the transduction of mechanical signals in the bone cells. It has been found that the mechano-transduction is mediated by Cx43 hemichannels. Interestingly, Cx43 hemichannels have been found active in the osteocytic cells, where they mediate fluid flow-induced PGE_2 and ATP release.

10.5   Connexins in the Inner Ear

Connexin distribution in the ear is mostly restricted to the inner part of the ear. Several connexins have been identified in the rodent ear, which include Cx26, Cx30, Cx31, Cx32, and Cx43. By analogy, most of these connexins have been identified in the human ear. Of these, Cx26 is the physiologically most important connexins found in the inner ear. In the inner ear, connexins are localized in the epithelia and connective tissue of the cochlea, thus connecting these tissues with the gap junction network. In addition, gap junction plaques and intercellular gap junction communication exists in the organ of Corti. Besides forming homotypic gap junction, heterotypic gap junctions, formed of Cx26 and Cx30, have been identified in the cochlear tissue. Gap junctional communication in the cochlear tissue is non-selective to ions, but there exists some sought of selectivity as far as passage of secondary messengers and other molecules are concerned. Besides intercellular gap junctions, functional hemichannels have been reported to exist in the organ of Corti. Hemichannels are involved in the uptake of large anionic molecules and under special circumstances release ATP to the extracellular space. The cochlea in the inner ear is the sensory organ that transmits sound signals. The cochlea contains many cells, which include epithelial cells, fibrocytes, and the sensory receptor cells called as hair cells. The cochlea has three compartments, namely scala media, scala tympani, and scala vestibule, and these are filled with two types of solutions. The scala tympani and scala vestibule contains perilymph, having ionic composition similar to that of extracellular solution, whereas the scala media contains endolymph, which possesses a high concentration of K^+ (150 mM). One of the important properties of endolymph is the high positive potential (+80 mV), termed as endocochlear potential. The endocochlear potential is produced by the stria vascularis, a two-layered epithelium forming the wall of the scala media. The cellular components of the stria vascularis contain the potassium channel Kir 4.1 in the plasma membrane of intermediate cells and K^+ transporters in the basal membrane of marginal cells of the. The circulation of K^+ ions from endolymph to perilymph is thought to be mediated by the gap junctional network between the supporting cells and epithelial cells on the basilar membrane and between the fibrocytes of spiral ligament and epithelial cells of stria vascularis. The importance of gap junction communication in the cochlea can be ascertained by the fact that its disruption leads to several forms of non-syndromic and syndromic deafness.

10.6   Connexins in the Endometrium

Gap junctional communication plays an important role for the proper functioning of the endometrium. Cx43 and Cx26 are the two major connexins which are expressed in the endometrium. Endometrium is highly dynamic in terms of its growth properties, having non-pregnant cyclic phases and a pregnant phase. Both Cx43 and Cx26 play an important role during the cyclic phases of non-pregnancy and during early pregnancy. Cx26 and Cx43 expression during the cyclic phases of non-pregnancy is mostly at the transcriptional level, which is confirmed by the increased levels of Cx43 and Cx26 mRNA, however with low amounts of the corresponding proteins. Just prior to the preimplantation stage, when the endometrium is ready to receive the embryo, the transcription of both connexins is downregulated. Maternal progesterone hormonal signal is regarded responsible for the transcriptional suppression of Cx43 and Cx26. Estradiol on the other hand upregulates the expression of Cx26 and Cx43. The decreased gap junctional communication during the early pregnancy is important for the differentiation of the receptive epithelium of the endometrium. Interestingly in rats and guinea pigs, the use of anti-progesterone drugs during the first days of pregnancy has been shown to inhibit embryo implantation. With the growing pregnancy, an intimate contact between embryo and the endometrium is required for the successful outcome of the pregnancy. During placental formation and its penetration into the endometrium, connexin expression is upregulated and gap junctional communication is established between the endometrial and the placental cells. This intercellular communication, mediated by the gap junctions, is important for the successful implantation and the placental invasion. During implantation, induction of gap junction connexins in the endometrium occurs in response to embryo recognition. The first connexin whose expression is induced is Cx26, and this results in the decidualization of the stromal cells surrounding the implantation chamber. The implantation of the blastocyst is accompanied by the expression of Cx43. Thus, spatial and temporal expression of endometrial Cx26 and Cx43, in response to embryo recognition, is important for the successful implantation of the embryo. Embryo recognition by the endometrium is mediated by different signals, which include hormones, secondary messengers, growth factors, prostaglandins, and the mechanical stimulations. Additionally, Cx43 expression and the gap junction communication in the blastomeres of the embryo are important for compaction. In rat embryo, Cx31 has been shown to have similar spatio-temporal expression as that of Cx43. In the initial stage, both Cx31 and Cx43 show even distribution in the inner cell mass and the trophectoderm. However, during the later stages, Cx31 show expression in the cells of the ectoplacental cone, which invades into the maternal decidual tissue, whereas Cx43 shows expression in the embryo proper. During the differentiation phase, Cx26 expression is highly induced in the labyrinthine trophoblast and is responsible for the feto-maternal exchange. With the maturation of placenta, expression of Cx31 and Cx43 decreases with increasing trophoblast differentiation.

10.7   Connexins in the β Cells of Pancreas

The pancreas constitutes an important gland of the vertebrates. It is both an endocrine gland, producing several important hormones, including insulin and glucagon, and a digestive organ, secreting pancreatic juice containing digestive enzymes that assist in the digestion of nutrients in the small intestine. The endocrine part of the pancreas is made up of clusters of cells called islets of Langerhans. In humans, millions of these cells are dispersed in the pancreas and constitute 1 % of total volume of the pancreas. The islets of Langerhans play an important role in glucose metabolism and regulation of blood glucose concentration. The islets of Langerhans contain many different cell types, each specific for different functions. For example, α cells secrete glucagon hormone in response to the low blood glucose level, β cells secrete insulin in response to high blood glucose level, and δ cells secrete somatostatin that regulates the function of α and β cells.

   β cells represent one of the important cell types of islets of Langerhans that perform crucial function of regulating blood glucose level by secreting an important hormone called as insulin. The clusters of the β cells require proper coordination for the secretion of insulin in response to high blood glucose level. This coordination integrates hundreds of β cells within each islet into a functionally homogeneous unit. β cells employ many mechanisms to achieve the proper coordination for the release of insulin. One of the important mechanisms for achieving the required coordination is the direct cell-to-cell coupling mediated by gap junction channels. Such channels help the cells to communicate with each other using various signalling molecules. The gap junction channels in the β cells are mostly made up of Cx36. Cx36 is a 321-amino acid protein with a long (99 amino acid) cytoplasmic loop containing an unusual stretch of 10 glycine residues and a short cytoplasmic COOH-terminal domain. The cytoplasmic loop and the carboxy-terminal domain of Cx36 contains potential recognition sites for various kinases, like casein kinases I and II, cAMP-dependent protein kinase, and calmodulin-dependent protein kinase II. Thus, Cx36 is a target of various kinases, consistent with the finding that, under certain conditions, the function of Cx36 is regulated by phosphorylation events. The gap junction channels formed by the Cx36 are mostly permeable to cationic species as compared to the negatively charged molecules. The importance of gap junction network between the β cells can be ascertained by the facts that the single cells (not in contact with other cells) show poor responsiveness to glucose stimulation, decreased basal expression of insulin, decreased pro-insulin biosynthesis, and less increase in the cytosolic calcium after glucose stimulation. However, the cells that are grown in contact with each other are very efficient in glucose responsiveness, insulin secretion, and cytosolic calcium increase. These observations are corroborated by using drugs that inhibit gap junction communication. For example, treating isolated islet cluster or the intact pancreas with carbenoxolone (gap junction inhibitor) resulted in decreased glucose responsiveness and insulin secretion, and these effects are reverted to normal when the carbenoxolone is washed away for the cells. Thus, the experimental evidence suggests that the gap junction channels between the cells are important for mediating glucose-induced secretion of insulin. Gap junctional communication between the cells allows them to equilibrate various ions and molecules between the cells for the coordinated function. This is what the gap junction channels do in the β cell physiology. The cell-to-cell communication mediated by connexins is regarded advantageous for the tissues made of different cell types. This is because the cells in a cluster show minor differences in their structure and hence are functionally asynchronous. It has been found that the differences in the biosynthetic activity of β cell would result in the irregular responsiveness to the glucose stimulation for the secretion of insulin. In other words, the release of insulin will be less and not synchronous in response to high glucose stimulation. Hence, the Cx36 gap junction channels between the β cells nullify the biochemical disparity and allow the synchronous response to the glucose stimulation. Under such conditions, β cell clusters release significantly larger amounts of insulin than the uncoupled cells. The inhibition of gap junction coupling between the β cells is known to alter the basal and glucose-stimulated insulin secretion, the expression of insulin genes, and the regulation of cytosolic calcium. The Cx36-null mice (Cx36−/−) have proved instrumental in discerning the role of Cx36 for the proper functioning of β cells. Lack of Cx36 results in the failure of calcium wave synchronization between the β cells. Consequently, loss of the synchronization affects the simultaneous release of insulin upon glucose stimulation. Thus, cell-to-cell contact between the β cells is critical for the glucose-induced insulin secretion. These and many other studies support the evidence of the importance of Cx36 signalling for the coordinate release of insulin by the β cells, and thus, the altered insulin secretion of Cx36-null mice results in abnormal control of blood glucose levels. Interestingly, the phenotypic effects of Cx36-null mice, like glucose intolerance, loss of circulating insulin oscillations, and increased β cell apoptosis, resemble to various β cell-related pathological phenotypes found in human beings. Since the human β cells are also coupled by the gap junctions formed by Cx36, it is safe to argue that various diabetes-related issues in humans are the consequence of altered Cx36 signalling. The significance of Cx36 in the β cell physiology is not limited to coordinate the insulin secretion in response to glucose stimulus, but also, Cx36 has been shown to protect the islets of Langerhans from the autoimmune attack mediated by various pro-inflammatory cytokines in the islet environment. The protection offered by Cx36 to the pancreatic cells is established from the events that occur during type-1 diabetes. In type-1 diabetes, the islets of Langerhans are self-attacked by various immunological factors, resulting in the reduced cell mass and hence the insufficient secretion of insulin. The role of Cx36 in providing protection against the self-attacking molecules has been demonstrated using transgenic mouse models. For example, transgenic mice over-expressing Cx36 significantly protected the β cells against cytotoxic drugs and cytokines that were shown to induce cell death similar to the onset of type-1 diabetes. Interestingly, mice lacking Cx36 showed increased sensitivity to these molecules and develop symptoms similar to type-1 diabetes. The molecular mechanism responsible for such protection, mediated by Cx36, is still a mystery. Various studies have shown that both gap junction-dependent and -independent mechanisms are responsible for the protection.

The role of Cx36 has also been reported in type 2 diabetes. Type 2 diabetes is a multifactorial disease, and association between type 2 diabetes and mutations in the chromosome regions that harbour Cx36 has been found. Although mutational analysis of Cx36 in type 2 diabetic patients has not been established in humans, mice possessing inactivated Cx36 depict certain diabetic phenotypes similar to what occurs during the early onset of type 2 diabetes in humans. The complete understanding of the role of Cx36 in diabetes will pave way for the innovative therapies in order to improve β cell functioning and hence blood glucose regulation.

Chapter 11
Connexin Functions

11.1 Connexins as Growth Regulators

Although connexins connect two cells by gap junctions, their expression in cells is known to regulate cell growth and differentiations. In fact, connexin expression shows an inverse relationship with the cell growth, and thus they have been aptly named as tumour suppressors. The role of connexins in regulating cell growth has been established from the observation that different types of tumour cells and tumorigenic cell lines, as well as solid tumours, show decreased or altered connexin expression and/or localization. Moreover, expression of connexins in the cancer cells is known to revert the oncogenicity of these cells, albeit to varying degrees. The role of connexins as tumour suppressors has been highlighted by the knockout animal models and knockdown strategies. It has been shown that Cx43-heterozygous and Cx32 knockout mice have increased susceptibility to tumorigenesis, whereas knockdown of the endogenous Cx43 expression, by small interfering RNA, resulted in a more aggressive tumour growth. In additions to the tumour growth, connexins also play an important role in the metastatic potential of the tumour cells. The extravasation of metastatic cells is dependent on the connexin expression. Thus there remains little doubt that connexins, in particular Cx43, Cx32, and Cx26, are crucial growth regulators in many types of cancer as well as in the normal cells. However, the mechanisms by which connexins and gap junction communication regulate growth are far from clear. This is likely due to many facets of connexin function and regulation. Connexins are known to regulate cell growth in a gap junction-dependent manner as well as gap junction-independent way. It is known that the carboxy-terminal domain of connexin is critical for the cell growth regulation. Connexin 43 is the most important player in regulating cell growth. Its C-terminal module hosts a number of phosphorylation sites and has the ability to interact with multiple proteins. As far as the channel-dependent role of connexins in regulating cell growth is concerned, it is believed that many specific and permeable metabolites or ions are responsible for growth control. Although many of these metabolites

© Springer India 2014
M.U. Hussain, *Connexins: The Gap Junction Proteins*, SpringerBriefs
in Biochemistry and Molecular Biology, DOI 10.1007/978-81-322-1919-4_11

are not known, however, many potential candidates, like Ca^{2+}, cAMP, and inositol triphosphate, have been identified. Besides the role of gap junctional communication in growth control, evidences are accumulating that connexins can inhibit growth in the absence of functional gap junction communication. For example, isolated cells, or use of gap junction uncoupling agents and prevention of membrane localization of the connexins, the regulatory effects of connexins on cell growth persist. Moreover, the C-terminal domain of Cx43, which lacks the capability of forming gap junction channels, is still able to inhibit cell growth as much as the full-length connexin. The molecular mechanism employed by the C-terminal domain of Cx43, in controlling cell growth, is still emerging. Association of C-terminal domain of Cx43 with many other proteins is regarded as one of the mechanisms responsible for the growth-regulating properties of Cx43. For example, protein tyrosine phosphatase is reported to bind to the carboxy-terminal of Cx43. Its association with Cx43 is expected to bring it to junctional complexes and enable it to dephosphorylate nearby proteins such as tyrosine kinase receptors or tyrosine-phosphorylated effectors of growth factor signalling. The second mechanism reflects the ability of cytosolic domains present in adhesion/junction proteins to interact with kinases, phosphatases, transcription factors, and structural proteins that are implicated in growth regulation. This interaction sequesters proteins from and prevents their signalling action on other possible targets at the cytosol and/or nucleus or at the cell surface. The C-terminal of Cx43 serves as an interactive platform for a variety of cellular proteins, like CCN3 with transcription factor properties. Several of these interacting partners are participants in the field of growth control; their interaction (and regulation of the interaction) with Cx43 or other connexins anchored to the plasma membrane may affect their subcellular localization and thus their site(s) of action. In one such study, it was demonstrated that the protein CCN3 interacts with the carboxy-terminal of Cx43. In the absence of Cx43, CCN3 is localized to the nucleus/cytosol; upregulation of CCN3 by Cx43 resulted in sub-membrane localization as well as secretion to the medium. Thus, Cx43 may inhibit growth by stimulating CCN3 expression as well as determining its subcellular localization.

Growth factors or oncogenes increase Cx43 phosphorylation on serine and/or tyrosine and generally decrease gap junction channel permeability, in a cell type-dependent manner. It has been demonstrated that growth inhibition by Cx43 appears to be inversely dependent on its phosphorylation at specific sites in response to growth factors or phorbol esters. The PKC target site S262, located at the C-terminal domain of Cx43, becomes phosphorylated in response to growth factor or tumour promoter stimulation in primary cardiomyocytes as well as in transformed HeLa and HEK293 cells. Simulating constitutive phosphorylation at S262 (in the mutant Cx43-S262D) completely abolished the ability of Cx43 to inhibit DNA synthesis in cardiomyocytes, without affecting its subcellular localization. Conversely, a Cx43 mutant, simulating lack of phosphorylation at that site (S262A), presented maximal growth inhibitory activity and enhanced dye permeability. Cx43 retained the ability to inhibit DNA synthesis in sparsely seeded myocytes, as has been shown for other cell types, indicating that cell–cell contact or gap junction formation per se are not required for the growth inhibition. It is plausible to argue that the cells in isolation are subject to different growth regulatory mechanisms than contact making cells and that the two models may not be comparable.

Cx43-mediated growth inhibition includes effects on secreted growth affecting molecules. Various studies have deciphered the role of the channel-dependent and -independent contribution of Cx43 in regulating cell growth. For example, U2OS cells expressing either intact Cx43 or CT-Cx43 displayed decreased cell proliferation as well as decreased levels of the S phase kinase associated protein 2 (Skp2). The SKp2 regulates the ubiquitination and thus the abundance of p27. By inhibiting Skp2, Cx43 and CT-Cx43 increase levels of the cell cycle inhibitor p27 in these cells. Furthermore, CT-Cx43 did not localize to plasma membrane at the cell–cell contact sites. This finding implied that Cx43-mediated growth inhibition is not dependent on the sub-membrane cellular milieu, cell–cell contact, and/or gap junction formation. Interestingly, CT-Cx43 was found to localize not only to the cytosol but also to the nucleus, suggesting the possibility for exerting effects on gene expression directly. Biological effects of CT-Cx43 in the heart under normal, non-ischaemic conditions suppressed cardiomyocyte DNA synthesis, adding primary cardiomyocytes to the list of cell types that are growth inhibited by CT-Cx43. In addition to inhibiting DNA synthesis, ectopically expressed CT-Cx43 can render myocytes more vulnerable to ischaemic stress. Nuclear localization of CT-Cx43 suggests the possibility for a direct role in gene expression relating to growth control and cell survival, by interacting, directly or indirectly, with, for example, chromatin and/or transcription factor complexes. A preliminary screen, using a mini array approach, indicated that ectopic expression of CT-Cx43 stimulated the expression of PIDD (p53-induced protein with a death domain) in cardiomyocytes. Expression of the anti-oncogene p53 was also found to be upregulated by CT-Cx43 in cardiomyocytes. It is well known that p53 acts as a tumour suppressor, inducing cell cycle arrest or apoptosis in response to a variety of stimuli; PIDD has been shown to function as a mediator of p53 effects in some systems.

11.2 Connexins in Cell Migration

In the multicellular organism, cell migration constitutes an important physiological phenomenon, so that each cell migrates to its proper location for performing specific functions. Cell migration is pivotal during embryonic development wherein cells migrate to specific locations for the proper development of multicellular organism. Errors during this process have serious consequences that include mental retardation, vascular disease, development defects, etc. In addition to its indispensible role during embryogenesis, cell migration equally contributes to the various physiological processes in the adult organism. For example, wound healing, angiogenesis, immune response, etc., require well-planned movement of cells in particular directions to localize specific regions. In addition, cell migration is also involved during various pathophysiological conditions, like cancer metastasis and vascular remodelling. The molecular mechanisms involved in the cell migration are far more complex than earlier thought. Cells use various chemical and mechanical signals to migrate to the specific locations. Besides the chemical and mechanical signals, many other protein factors are involved in the specific migration of the cells. Most of these proteins belong to cytoskeleton protein family. In addition, it has been demonstrated that the gap junction proteins, connexins, are also important players in cell migration. The involvement of connexins in cell migration has been confirmed using in vitro and in vivo experimental approaches. In most of the cell types, connexins have been shown to be important for cell migration. It is well established that connexins help in the migration of brain cells, like astrocytes, endothelial progenitor cells and endothelial cells during wound healing, and cardiac cells during the heart development. Of all the connexins, the role of Cx43 in cell migration has been established unequivocally. The Cx43 mediate angiotensin II (AngII)-induced cell migration of vascular smooth muscle cells. In C6 glioma cells, expression of Cx43 induces cell motility and tissue invasivity. Thus, there is no doubt for the involvement of connexins in the cell migration; however, the molecular mechanisms responsible for such role are still not well understood. Moreover, it remains to be established whether cell migration mediated by connexins is channel dependent or independent. Additional complexity is added by the data indicating the involvement of connexin hemichannels in cell migration. Although studies have shown the role of gap junction channels in the cell migration, most of the data favour the involvement channel-independent role of connexins in cell migration. In the following paragraphs, the role of channel-dependent and -independent mechanism will be discussed.

11.2.1   Channel-Dependent Role of Connexins in Cell Migration

Channel-dependent role of connexins in cell migration requires functional gap junctional channels between the adjacent cells. For the coordinative and collective migration of cluster of cells, for example, during embryogenesis, communication network between the migrating cells is an essential feature. This is provided by the gap junction communication mediated by the gap junction channels. The requirement of the functional gap junction communication for cell migration has been demonstrated in wound healing. During wound healing, the polarization and migration of smooth muscle cells is mediated by the increased levels of calcium in these cells. For the coordination of cell migration, the calcium wave is to be propagated between the clusters of migrating cells, and this is achieved with the help of functional gap junction channels. These studies are endorsed by the dominant negative mutants of Cx43 that form non-functional gap junction channels and thus are incapable of inducing migration of the cells. In addition to the role of gap junction channels between the cells, functional hemichannels have also been implicated to play a role in cell migration. Through the functional hemichannels, ATP is released in the extracellular space, and this helps in generating calcium waves in the neighbouring cells and thus induces cellular migration.

11.2.2 Channel-Independent Role of Connexin in Cell Migration

Although the requirement of functional gap junction channel for cell migration is immense, there are evidences which suggest the role of channel-independent function of connexin in cell migration. These observations are based on certain experimental data wherein the cell migratory property of carboxy-terminal truncated Cx43 is significantly compromised, even though the mutant connexins were able to form functional gap junction channels. Similarly, carboxy-terminal truncated Cx43 transgenic mice demonstrated defective neuronal cell migration and development of brain. Further studies have shown that the neural migration depends upon the adhesive properties of gap junctions rather than on their channel properties. In one such study, it was established that during cerebral cortex development, the newly formed neurons migrate to the cortical plate along the radial glial fibres. The molecules that mediate this movement were found to be Cx26 and Cx43. Although, dominant negative mutants of these connexins resulted in the formation of closed channels, the connexins retained the property of inducing neuronal migration. Further investigation revealed that the channel defective connexins retained the adhesion properties and that leads to the neuronal migration. Mutational studies indicated that the amino acids present on the external loop of Cx43 in the neural and glial cells are directly involved in providing the required adhesion for cell migration. These and other studies proved that the gap junctions are necessary for neural migration in the developing cortex and that function is based on adhesive property rather than channel activity of gap junctions. Additional studies have further confirmed the connexin-mediated cell adhesion is required for migration. For example, during brain injuries the microglia rapidly migrates towards the wound, accompanied with high Cx43 expression. Thus, it is established that the Cx43-mediated cell adhesion is required for the migration of microglia. The adhesion-based cell migration is thought to involve the interaction of connexins with various extracellular matrix proteins. This interaction is regarded essential for the downstream intracellular signalling.

Besides the adhesion and channel-mediated cell migratory role of connexins, intracellular signalling has also been found critical for the cell migration. These are based on the observations that connexin can also mediate single cell migration, which is unlikely to be mediated by connexin adhesive guidance or gap junction communication with other cells. The intracellular signalling is thought to be mediated by the interaction of connexins with various protein or signalling molecules. Some of these interactions are known to stabilize the membrane proteins, like N-Cadherin and ZO-1 that are involved in the cell migration. The carboxy-terminal domain of connexins is the site for various interactions that are crucial for the cell migration. The importance of carboxy-terminal domain in cell migration can be ascertained by the fact that the carboxy-terminal truncated Cx43 fails to induce cell migration. Additionally, ectopic expression of carboxy-terminal domain of Cx43 has been shown to retain the capability of inducing cell migration. Although, the role of connexins in cell migration is pretty well established, the mechanistic details are still in infancy. Further studies are required to elucidate the molecular interaction and signalling pathways that make the cell to migrate to a specific location.

11.3   Connexins and Gene Expression

It is becoming increasingly clear that connexins have profound effects on gene expression. Gene array analyses of transcription profiles of tissues and cells from specific connexin-deficient mice have indicated that there are large-scale alterations in these connexin null transcriptomes, from which it is inferred that gap junction genes may be "hubs" in gene expression networks. Therefore, in addition to the role of gap junctional communication in the maintenance of homeostasis, morphogenesis, cell differentiation, and growth control in multicellular organisms, altered phenotype in connexin-deficient mice and in disease-causing mutants could also arise from altered gene expression. cDNA array data has been used to explicitly delineate the roles of Cx32 and Cx43 in the expression of growth and development genes in brain and heart. The probability of growth genes being up- or downregulated in these connexin-deficient organs is significantly higher than that for genes in other categories, indicating that expression of growth genes is more dependent on the expression of Cx43 and Cx32 than is the average gene. The growth genes which are up- or downregulated in both Cx43 null brain and heart include Crkl (which plays a specific role in integrin-induced migration as a downstream mediator of Src), Elk4 (a member of ETS oncogene family with transcription factor activity), Rasa1 (allowing the inactivation of the anti-apoptotic function of N-terminal fragment), and Vegfa (involved in angiogenesis). This predictability extends to growth genes, thereby adding new arguments in favour of connexin implication in the growth control. Moreover, comparison of genes regulated in Cx43 and Cx32 null brains indicates both a remarkable degree of overlap (which may be due either to channel-mediated effects or to common binding of regulatory molecules to both connexins) and also connexin-dependent transcriptomic effects (which could be due either to permeability differences between these connexins or to different binding partners).

11.4   Connexins in Shear Stress

The role of connexins in shear stress is mainly involved with the normal functioning of vascular endothelium. The vascular endothelium is a thin monolayer of cells that line the luminal side of all blood vessels. It acts as a barrier between the blood and the surrounding tissue and is involved in the exchange of fluid, electrolytes, macromolecules, and cells between the intravascular space and surrounding tissue. The endothelium is always braced with haemodynamic shear stresses due to the flow of blood at the luminal surface of blood vessel. The endothelium is highly dynamic in nature and has evolved a number of mechanisms to sense shear stress. Various cell membrane receptors, membrane channels, adhesion molecules, cytoskeleton, etc., are regarded the molecular candidates that sense the shear stress in endothelial cells. The shear stress forces constitute important signalling mechanisms that regulate most aspects of vascular physiology. Thus, these forces are crucial for the proper development and functioning of the vasculature. These facts can be ascertained by various arterial diseases that are nucleated by the deregulation of endothelial cells to the shear stress.

The connexins are regarded important molecule players that regulate the functioning of endothelial cells and are regarded important sensors of sheer stress. The endothelial cells are known to express Cx37 and Cx40. In addition, some of the endothelial cells, at regions of high fluid turbulence, for example, branching points of arteries, are known to over-express Cx43. However, Cx37 is regarded as a main connexin that is involved in sheer stress in the endothelial cells. Endothelial cells (ECs) of healthy arteries express high levels of Cx37. High laminar shear stress (HLSS) is known to induce the expression of Cx37 and thus mediates vasculo-protective effect. The effect of shear stress on Cx37 expression participates in the overall protective effect of high laminar flow on the endothelium. The over-expression of Cx37 upon high laminar shear stress is mediated through the induction of Kruppel-like factor 2 (KLF2). KLF2 constitute an important transcription factor that is known to play a critical role in the differentiation and function in a variety of cell types, including T lymphocytes, adipocytes, and lung cells. Moreover, KLF2 possess an anti-proliferative and pro-survival factor, properties that are identical to the effect of fluid shear stress on endothelial cells. Interestingly, the promoter region of the Cx37 gene is known to harbour KLF consensus binding sites that upregulate the transcription of Cx37 gene. High laminar shear stress shows a strong correlation between the increased KLF2 and Cx37 expression. Further confirmation on the role of KLF2 in inducing Cx37 transcription came from the studies depicting that the silencing of KLF2 transcription factors leads to the drastic downregulation of Cx37 expression. Thus, high laminar shear upregulates the expression of KLF2 and concomitantly that of Cx37. Therefore, the effect of shear stress on Cx37 expression may contribute to the synchronization of endothelial cells and hence participate in the protective effect of high laminar shear stress.

The role of Cx37 in the endothelial cell physiology is further explained by certain arterial diseases, like atherosclerosis. The expression of Cx37 expression is lost in the endothelial cells overlying atherosclerotic plaques. The deletion of Cx37 in ApoE−/− mice increases their susceptibility to atherosclerosis, which suggests that Cx37 has anti-atherogenic properties. Besides Cx37, the expression of Cx40, Cx43 is known to change during atherosclerosis.

11.5 Hemichannel-Dependent Functions of Connexins

Apart from gap junction communication, connexin-forming hemichannels (HCs) play critical roles in modulating cellular functions. The hemichannels allow the transfer of small molecules that regulates various cellular functions. The molecules released through the open hemichannels bind to cell surface receptors and lead to the activation of intracellular signalling pathways. Consequently, the activation of signalling pathways leads to the regulation of cellular function and physiology through gene transcription and translational or post-translational events including intracellular trafficking, protein turnover, phosphorylation, and other modifications. The hemichannels formed by different connexins may regulate distinctive cellular events via the selective passage of specific molecules. Some of these cellular events are discussed below.

11.5.1   Cell Cycle

The role of connexin hemichannels in the regulation of cell cycle, cell progression, and development is still unclear. Certain human diseases advocate the significance of the hemichannel-mediated function of connexins. For example, human Cx26 is a susceptible locus in patients with psoriasis, a skin disease that occurs due to increased skin cell division. Cx26 is known to form hemichannels in the skin and inner ear, and mutations of Cx26 genes are associated with skin diseases and deafness. However, it is not clear if abnormally elevated cell division in psoriasis is related to the function of Cx26 hemichannels. In C6 glioma cell, Cx43 is shown to reduce cell proliferation by impeding the cell cycle progression from G0/G1 to S phase. Extracellular Ca^{2+} levels and pH, which regulate hemichannels opening, are also known to regulate cell growth and proliferation. Cx43 hemichannels in 3T3 fibroblasts release NAD^+, which is converted to cyclic ADP-ribose (cADPR) by an ectoenzyme, CD38. Uptake of cADPR by the cell increases intracellular calcium levels, consequently increasing cell proliferation rate by shortening S phase. Neural precursor cells release ATP, and released ATP controls intracellular calcium levels via Cx43 hemichannels. Moreover, calcium waves, propagating through the radial glial cells, require connexin hemichannels, and disruption of the wave decreases the proliferation of the cortical ventricular zone (VZ). Blocking of hemichannels induces phosphorylation of small GTPase cdc42, which is important in orchestrating cytoskeleton organization during cell division. Gap27 peptide, which is designed to block Cx43 hemichannels, hinders T-cell proliferation by increasing the G_0/G_1 cell numbers, suggesting the role of hemichannels in sustaining T-cell clonal expansion.

During chick retinal pigment epithelium (RPE) development, robust Ca^{2+} waves are generated, and gap junction coupling is involved in this process. Specific peptide blocker, Gap26, of Cx43 hemichannel inhibits release of both ATP and Ca^{2+} waves in RPE. The ATP released through Cx43 hemichannels expressed in RPE influences the mitotic division in retinal ventricular zone, which is important in generation of neurons and glia. Extracellular ATP is a signal for various events during embryonic development. ATP through purinergic receptors evokes Ca^{2+} waves and promotes retinal progenitor cell proliferation. ATP signalling also initiates the proliferation of rat cerebral cortical astrocytes, human neural stem cell, and Müller glial cell.

Hence, the release of ATP through hemichannels activates purinergic receptors and triggers gene expression necessary for retinal development, ionic homeostasis in cochlea, and proliferation of neural precursor cells during neural development.

11.5.2 Ischaemic Preconditioning

A subtle or sublethal insult to a tissue causes protection of the tissue towards a severe insult, and the process is known as preconditioning. There is a general belief that release of protective agents during the initial insult leads to resistance towards permanent tissue damage. Ischaemia could be caused due to deprivation of oxygen and glucose or ATP efflux. Cardiac and the neural tissues are most extensively studied to understand the preconditioning phenomenon, and various connexin hemichannels are implicated in the release of tissue protective agents. Under ischaemic stress, cardiac myocytes and C6 glioma cells release ATP. ATP efflux leads to increase in intracellular calcium levels; consequently, hemichannels open further causing an increase in cellular volume, which ultimately results in apoptotic cell death. Cx43 hemichannel blocking peptide, Gap26, is demonstrated to protect rat neonatal cardiomyocytes and intact rat heart from ischaemic injury. Cardiac myocytes subjected to ischaemia showed an increase in release of ATP, which can be blocked by Gap26 peptide. Moreover, ATP release through Cx43 hemichannels is reduced due to prolonged ischaemia. Regulation of hemichannel opening during ischaemia could be important in preventing necrosis due to sustained ATP release. Extracellular ATP is known to communicate through G-protein-coupled P2Y2 and P2Y4 purinergic receptors that are expressed on the cell surface and invoke intracellular calcium levels. Other factors released by hemichannels and the signalling pathways triggered during preconditioning response are yet to be determined. Apart from the connexin at the cell surface, recent studies have demonstrated that Cx43 is also expressed in mitochondria and plays an important role during ischaemic preconditioning. Mitochondria are known to be important for mediating precondition response. Treatment with gap junction and hemichannel blockers, such as carbenoxolone and heptanol, results in a decrease of mitochondrial dye uptake, implying the existence of functional hemichannels in the mitochondria. During prolonged ischaemia, ATP levels decrease, leading to lowering of pH and accumulation of reactive oxygen species (ROS), which ultimately results in cell death. These studies suggest that connexin hemichannels play an important role in the release of ATP during the early stages of cardiac ischaemia, thereby creating a protective preconditioning effect. Additionally, Cx36 is abundantly expressed in neuronal tissue and Cx43 is localized to glial tissue. Cx43 and Cx36 hemichannels show increased opening during ischaemia leading to the release of ATP. Stimulation by acute ischaemic conditions enhances neuronal Cx36 and glial Cx43 hemichannel activity, which favours ATP release and generates preconditioning. Besides ATP, Cx43 hemichannels also mediate the release of glutamate from astrocytes and glutathione and other amino acid derivates from hippocampal slices. The efflux of glutathione is postulated to have cell protective functions in situations when glutamate reuptake is impaired, such as after stroke. Similar to the observations made in the cardiac tissue, acute ischaemia opens hemichannels, leading to ATP release and depletion of intracellular ATP, which finally results in cell death. Metabolic inhibition is also reported to increase ATP release via hemichannels, and the depletion of intracellular ATP reserves could promote cell death. Metabolic inhibition also increases Cx43 expression on the cell surface causing further imbalance in the cellular ionic concentrations, thereby, making the cells vulnerable to cell death. However, Cx36 hemichannels expressed in the neurons are observed to mediate ischaemic pre-conditioning. Cx36 hemichannels in neurons mediate the ATP release and cause depolarization of the neurons. Depolarization of neurons ultimately increases neuronal tolerance towards ischaemia. The role of hemichannels in causing glutamate toxicity due to the release of excessive glutamate during ischaemic conditions is yet to be studied. Overall, hemichannels formed by connexins promote cell death in both neurons and cardiomyocytes during acute ischaemia, but they also can promote tolerance towards ischaemia after an encounter with sublethal ischaemia.

11.5.3   Mechano-transduction

Expression of connexins is observed to increase after mechanical stimulation of cardiac and bone cells. Gap junctions present in these cells are known to be important conduits of signalling molecules that are generated due to mechanical stimulation, indicating that connexins act as mechano-sensory channels. In osteoblasts, both Cx43 and Cx45 are expressed. Cx43 knockout mice displayed defective skeletal development and delayed ossification, suggesting that Cx43 is important in adaptation to mechanical stimulation. These observations indicate that Cx43 is regulated by mechanical stimulation and is the major gap junction-forming protein in osteoblasts and osteocytes. Moreover, mechanical stimulation leads to opening of hemichannels in many cell types. In human osteoblast cells and osteoblast-like MC3T3-E1 cells, mechanical stimulation leads to the release of IP3, Ca^{2+}, and ATP through Cx43 hemichannels. Cx43 hemichannels expressed in osteocytes are involved in the release of bone regulators, such as PGE_2 and ATP in response to fluid flow shear stress. Other bone cells, such as chondrocytes, also express Cx43 hemichannels, which are involved in the release of ATP in response to cyclic compressive strains. In osteocytes, the opening of Cx43 hemichannels is adaptively regulated in response to the magnitudes and durations of mechanical stimulation. Cx43 hemichannels are known to be important for cell survival, which is critical in maintaining the integrity of bone. Osteosarcoma cells when treated with drugs that induce cell cycle arrest resulted in expression of functional hemichannels and formation of dendritic processes similar to osteocytes, implying that hemichannels may be involved in osteoblast differentiation and osteocyte survival.

Chapter 12
Connexin Mutations and Disease

Since connexins are ubiquitously expressed in mammalians and appear to have a unique role in cell and tissue homeostasis, it is expected that mutations in these genes would be associated with a wide variety of diseases. In fact, since the first description of a mutation in the Cx32 in patients with the X-linked Charcot–Marie–Tooth (CMTX) syndrome, disease-causing mutations have been detected in almost every connexin gene. Mutations in a connexin can affect one organ expressing this protein but simultaneously spare another organ expressing the same connexin. This selectivity for organ involvement can partially be explained by redundancy in connexin expression, especially in cells where connexins are indispensable to survival of the organism. It is also quite surprising that some connexin genes harbour hundreds of different mutations in humans whereas in others no mutations have been found so far. This difference in susceptibility to mutagenesis is poorly understood. As connexins have similar gene structure and a high sequence homology, it seems highly unlikely that only some of these genes would contain mutational hotspots, and we can also assume that the mutation rate is comparable in each connexin gene. If mutations in one particular gene would have little consequence on proper cell function, then numerous gene polymorphisms should be found in the human population. This is, however, not the case for connexins, as only a limited number of polymorphisms have been described in screenings of those genes. Some of the mutations in connexin genes and their linkage to diseases are discussed below.

12.1   Cx43 and Oculodentodigital Dysplasia (ODDD)

Cx43 is the most widely expressed member of the connexin protein family. It is therefore not surprising that mutations in the Cx43 gene are associated with a broad spectrum of phenotypic alterations. In fact, mutations of this gene have been tightly correlated to the oculodentodigital dysplasia (ODDD) syndrome, with over

© Springer India 2014

M.U. Hussain, *Connexins: The Gap Junction Proteins*, SpringerBriefs
in Biochemistry and Molecular Biology, DOI 10.1007/978-81-322-1919-4_12

60 mutations described so far. This rare, usually autosomal dominant syndrome is characterized by (1) striking faces with a long, thin nose with hypoplastic alae nasi, anteverted nostrils, and a prominent bridge; (2) ocular involvement including microphthalmia and/or microcornea; (3) dental anomalies (enamel dysplasia and its consequences); and (4) digital manifestations, usually involving syndactyly of the fourth and fifth fingers as well as their deformations. Other reported skeletal anomalies include cranial hyperostosis and broad tubular bones. About 30 % of patients also present with neurological symptoms (lower body weakness and spasticity, gait disturbances, lack of bladder or bowel control, and white matter changes). An increasing incidence of conductive hearing loss, hyperkeratotic symptoms, and rare cardiac anomalies (atrioseptal defects and arrhythmias) has also been reported. The majority of Cx43 mutations described in relation to ODDD are dominant missense mutations, located in the first two-thirds of the Cx43 protein. Interestingly, only two mutations have been detected in the CT, both of which induce a frameshift followed by an early truncation of the protein (Y230fsX236 and C260fsX306). Mutations are equally distributed over the majority of the Cx43 protein (except for the CT), and there is a high inter-patient phenotypic variability, even within families carrying a single mutation. However, patients suffering from recessive mutations generally had more severe phenotypes, as often seen with autosomal recessive conditions. One exception to this lack of genotype–phenotype association may be the association of both CT frameshifts and truncation with dermatological symptoms, as both mutations were found in patients displaying palmar or palmoplantar keratoderma (PPK). As a mouse mutant lacking the majority of the C-terminal domain of Cx43 (M257X) has defects of the epidermal water barrier, it is tempting to speculate that this domain plays a particular role in skin homeostasis. As an attempt to better understand the peculiar manifestations of Cx43 mutations, 14 of these described mutations have been investigated for functional effects at the cellular level. All examined mutants, when expressed in cell systems, appear to have a reduced or abolished capacity to form functional gap junctions, despite a normal localization for most of them. Moreover, the mutants studied in combination with wild-type Cx43 (G21R, Q49K, L90V, G138R, R202L, and V216L) have a dominant negative effect on endogenous or transgenic Cx43 gap junctions. Hemichannel activity has also been suggested to play a role in the pathogenesis of ODDD. As the ODDD phenotype is characterized by craniofacial deformities and additional long bone alterations such as syndactylies and tubular bones, a more specific role for Cx43 in osteoblast differentiation has been investigated. Whereas transfection of Cx43 mutants in committed osteoblasts had little consequence, the use of a mouse model for ODDD (having a heterozygous Cx43 G60S mutation) suggests that Cx43 mutations in the germ line, and thus present in early bone development stages, could affect late stages of osteoblast differentiation. In fact, the presence of a germ line Cx43 G60S mutant appeared to reduce the phosphorylation state of wild-type Cx43 in osteoblasts, to reduce the formation of gap junction channels, and subsequently to delay the differentiation of these osteoblasts.

12.2 Cx46, Cx50, and Cataract

The lens contains two different cell types: epithelial cells on the anterior surface, which express Cx43 and Cx50, and specialized lens fibres expressing Cx46 and Cx50. Congenital cataracts are a very rare group of diseases (1–6 cases per 10,000 live births) of which about 25 % are inherited, usually in a non-syndromic autosomal dominant pattern. Mutations in both Cx46, whose expression is restricted to specialized lens fibres, and Cx50, a gene expressed throughout the lens, have been associated with zonular pulverulent cataracts, characterized by numerous powdery or punctate opacities in the lamellar regions surrounding the embryonic nucleus. All Cx46 missense mutations described so far are located in the N-terminal domain (D3Y, L11S), the first transmembrane helix (V28M, F32L), or the two ELs (P59L, N63S, R76G, R76L, P187L, and N188T). Notably, the L11S mutation is characterized by a distinct phenotype where very dense structures are embedded in the perinuclear layers of the lens. Another Cx46 mutation, leading to a frameshift and to a sequence of 87 aberrant amino acids after codon 379 in the CT, has been investigated in more details. This mutant is unable to form gap junctions, because of the presence of a diphenylalanine retention signal in the aberrant CT, which retains the mutant protein in the ER and impairs proper trafficking to the membrane. Mutations in Cx50 are located in the NT (R23T), the first EL (G46V, E48K, S50P), the second transmembrane domain (P88S, P88Q), or the CT (I247M). Several of those Cx50 mutants (R23T, S50P, P88S, and P88Q) are shown to be unable to form gap junctions in transfected cell systems, because of an intracellular retention of the mutant protein which appeared to impair trafficking of normal co-expressed Cx50. Interestingly, the P88S mutant, which is associated with a severe phenotype of nuclear pulverulent cataract, was shown to form cytoplasmic inclusions because of a reduced degradation. However, the P88Q mutant, which was retained in the ER/Golgi and only had a partial dominant negative effect on normal Cx50, was detected in patients with lamellar, but not nuclear pulverulent, cataract. It is thus conceivable that a partially conserved gap junctional intercellular communication in the heterozygous P88Q mutant was able to ensure sufficient function at least during embryonic development of the lens. Apart from trafficking defects, two other mechanisms have also been suggested for Cx50-associated cataract. First, the G46V mutant can elicit normal gap junctional coupling between cells. Nevertheless, it was found to display enhanced hemichannel function. As the aqueous humour has a relatively low Ca^{2+} concentration, these hemichannels may be functional and lead to cell damage or death by ionic and metabolic leak. Secondly, although the Cx50 S50P mutant only slightly modified electrical properties of Cx50 gap junctions, it was shown to have a dramatic effect on Cx43 gap junctional communication in transfected cells and in mouse lens epithelial cells. This suggests that gap junctions in the lens epithelial cells may play an additional role in maintaining lens transparency. In mice, knock-out of Cx50 resulted in significantly reduced lens and eye growth in addition to mild nuclear cataract. By contrast, deletion of Cx46 produced severe cataracts resulting from the failure to maintain crystallin solubility but did not alter ocular growth. Functional replacement of Cx50 by Cx46 (Cx50KI46 mice) prevented the loss of crystallin solubility and cataracts that occurred following knock-out of the Cx50 gene, but the growth deficiency remained unchanged. Nowadays, the lens is still considered as a classical example of connexin isotype-specific segregation of the contributions of gap junction communication to function, that is, to the control of normal growth and to the maintenance of clarity.

12.3   Cx32, Cx47, and Myelin-Related Diseases

Cx32 is expressed in myelinating Schwann cells, whereas Cx47 is mainly found in oligodendrocytes. In fact, mutations in those two connexin genes, respectively, lead to myelin diseases of the peripheral nerves (for Cx32) and of the central nervous system (for Cx47). Charcot–Marie–Tooth disease (CMT) is a heterogeneous group of diseases characterized by progressive peripheral nerve degeneration. Patients typically present with distal muscle weakness, amyotrophy, decreased tendon reflexes, and peripheral sensory loss. Several mutations in genes for peripheral myelin proteins have been described, although the X-linked form of CMT (CMTX) is caused by mutations of the Cx32 gene. Cx32 is expressed in non-compact myelin in the paranodal regions and forms gap junctions between the folds of Schwann cell plasma membrane. As these gap junction channels considerably shorten the diffusion time across the concentric layers of the Schwann cell, it is hypothesized that Cx32 gap junctions may participate in the transfer of metabolites and other small molecules through the myelin sheath and to the axon. Other connexins are also expressed in non-compact myelin, as evidenced by studies on Cx32 KO Schwann cells, and are expected to play a role in the communication between the folds of Schwann cells. To date, more than 200 different mutations in the Cx32 gene have been described in relation to CMTX. The effects of a consequent number of these mutations have been investigated at the cellular level. Most of these mutations result in a loss of function through the absence of formation of gap junction plaques or alter the gating properties of Cx32 gap junctions without affecting expression or conductance. Indeed, mutations located in the CL (E102G, del111-116) or in the CT (Y211stop, R220stop) cause an increased sensitivity to voltage or cytoplasmic acidification. Whereas voltage-dependent gating is not expected to play a major role in vivo (as the gap junctions are formed between two layers of the same cell), a more prominent closure of gap junctions in response to low pH could result in a functional alteration in conditions of ischemia or metabolic stress. Interestingly, one of these CL mutations affecting channel gating (E102G) was detected in a patient harbouring a milder phenotype of CMTX when compared with patients with mutations having a more drastic effect on Cx32 function. On the other hand, it has also been demonstrated that some Cx32 mutations lead to a toxic gain of function. Two missense mutations (S85C in TM2 and F235C in the CT) are shown to increase hemichannel permeability, leading to a potential loss of small metabolites and of ionic gradients as well as to an influx of Ca^{2+}. In fact, the transfection of these mutant forms in oocytes led to rapid cell death. Therefore, some Cx32 mutations may lead to direct Schwann cell toxicity, which correlates well with the demyelination shown in histological samples of patient nerve biopsies.

Pelizaeus–Merzbacher disease (PMD) is a hypomyelinating disorder of the central nervous system. It is clinically characterized by nystagmus, progressive spasticity, ataxia, developmental delay, and dysarthria. The diagnosis is confirmed by magnetic resonance imaging, showing a characteristic uniformly increased intensity of the white matter signal in T2-weighted images (because of the reduction in myelin content and replacement by aqueous substance). PMD is caused by muta-tion or duplication of the PLP1 gene that codes for proteolipidprotein1, the main component of CNS myelin, and presents as a recessive X-linked disease. Patients having the same symptomatology but lacking PLP1 alterations are considered as having Pelizaeus–Merzbacher-like disease (PMLD). A varying proportion (between 7 and 50 %) of these patients was shown to have mutations in the Cx47 gene. At least 20 different mutations have been reported so far, spanning the entire coding sequence and causing missense mutations, nonsense mutations, or frameshifts in the coding sequence. Using HeLa transfectants, it has been shown that all three Cx47 mutants have an impaired trafficking to the membrane and are mainly retained in the ER. Assays evaluating gap junctional permeability or conductivity could not detect any coupling between these mutant-bearing cells. I33M, a missense mutation located in the first transmembrane domain of Cx47, was found to cause a milder form of PMLD with a slower progression, mainly defined by a spastic paraplegia but lacking the characteristic nystagmus. Interestingly, when this mutated form of Cx47 was transfected into HeLa cells, it was not retained in the ER but could form gap junction plaques at the cell–cell interface. Similarly, to the other mutants, however, Cx47 I33M could not form functional channels. As this particular mutant was associated with a milder form of disease, it is speculated that the presence of Cx47 at the plasma membrane is important for oligodendrocyte function, even in the absence of functional gap junction channels. This may point towards a role of hemichannels or of protein–protein interaction.

12.4   Cx26, Cx31, and Cx30 Genes and Skin Diseases and Hearing Malfunctions

It has been shown that mutations in connexins Cx26, Cx30, Cx30Æ3, and Cx31 can cause sensorineural deafness, alone or in combination with hyperproliferative skin disorders. Those four connexins have been detected in the inner ear cochlea and are thought to play a central role in potassium recycling to the endolymph. The autosomal recessive inheritance pattern predominates in non-syndromic genetic deafness (80 %), the rest following mostly an autosomal dominant pattern. The Cx26 gene is the most important single locus linked with recessive non-syndromic sensorineural hearing loss, with 10–50 % of all patients bearing mutations in this gene, depending on the studied population. Interestingly, most Cx26 mutations associated with a recessive pattern of deafness cause either premature truncation of the protein or missense mutations in the last two-thirds of the protein. For instance, the 35delG mutation, which accounts for the majority of Cx26 mutations, especially in Caucasian populations, drives a deletion of one of six guanines at positions 30–35, leading to a frameshift and a premature stop codon at nucleotide 38. Overall, mutations causing an early truncation cause a profound hearing loss, whereas missense mutations have a more variable effect. Of those, mutated Cx26 proteins unable to form gap junctions in transfected cells (W77R, S113R, R143W, V153I, M163V, R184P, L214P, etc.) are associated with a profound hearing loss in the homozygous state. By contrast, it has been shown that patients bearing the N206S missense mutation have only mild to moderate hearing loss. Interestingly, this particular mutation does not impair the formation of functional gap junctions but reduces conductance levels and alters gating properties. Another mutation associated with profound deafness, V84L, is of particular interest because its electrical properties are virtually unaltered. However, it has been shown that the permeability of the mutated channel to IP is dramatically reduced, leading to a block in Ca^{2+} wave propagation in the supporting cells of the organ of Corti. It is thus proposed that metabolite transfer between cells, and not electrical coupling, is the determinant role of gap junctions for potassium homeostasis of the inner ear. By contrast, the Cx26 mutations associated with the less common autosomal dominant forms of non-syndromic deafness, or with syndromic deafness, are located in the TM1 or the EL1. The mutations investigated in cell systems all showed a dominant negative effect on non-mutated Cx26, as well as a trans-dominant effect on other connexin isotypes, whether the mutated form had impaired trafficking (D66H) or formed non-functional gap junction plaques (M34T, G59A, or R75W). There thus seems to be a strong correlation between a dominant negative effect of a mutated connexin on normal connexins and the observed pattern of dominant inheritance. Mutations in the other cochlear connexins have been less extensively studied. It is has been shown that digenic inheritance of recessive deafness by mutations in Cx26 and Cx30 or in GJB2 and Cx31 can occur. In other words, deafness can be caused by the addition of a mutation in one allele of Cx26 and one allele of Cx30 or Cx31, indicating an interaction of these connexins in the cochlea. Most connexin-associated skin disorders, whether they segregate with hearing impairment or occur alone, are inherited in an autosomal dominant pattern, which points towards mutations causing a toxic gain of function or a dominant negative effect on non-mutated connexins.

Cx30 mainly parallels the distribution of Cx26. Cx30Æ3 and Cx31 are expressed during later stages of epidermal differentiation. These expression patterns fit well with the skin disorders associated with mutations in each connexin. Cx26-associated syndromes have a hyperproliferative component and characteristically include a PPK and deafness associated with nail and hair dystrophy, keratitis, and a predisposition for squamous cell carcinoma (keratitis–ichthyosis–deafness syndrome, or KID); keratoderma on joints and constrictive bands on digits (Vohwinkel syndrome); or knuckle pads and leuconychia (Bart–Pumphrey syndrome). Similarly, Cx30 mutations can cause KID but also a distinct though overlapping phenotype (Clouston syndrome) defined by PPK, alopecia or hypotrichosis, and nail dystrophy. On the other hand, Cx30Æ3 and Cx31 mutations affect more keratinocyte differentiation than proliferation. The main syndrome caused by mutations in these genes is erythrokeratodermia variabilis (EKV), characterized by migrating erythematous patches on the trunk and a fixed, usually symmetrical keratoderma without palmo-plantar involvement. Intriguingly, the vast majority of mutations found to cause skin disorders are located in the NT, the TMs, or EL1, which indicates a crucial role of these domains for connexin function in the skin. The properties of several specific mutations have been investigated in more detail. A recurrent finding, especially for mutations in Cx26 causing KID (such as G12R, A40V, G45E, and D50N) and Cx30 mutations associated with Clouston syndrome (G11R and A88V), was that these alterations did not appear to impair the formation of gap junction plaques and only mildly affected gap junction channel properties. However, all these mutations were shown to increase hemichannel function. Interestingly, HeLa cells transfected with Cx30 mutants (G11R or A88V) were shown to release increasing amounts of ATP through their mutant hemichannels. As ATP, by activating specific purinergic receptors, is known to regulate keratinocyte proliferation and differentiation, it is conceivable that Cx30 (and, by homology, Cx26) mutations cause hyperproliferative skin diseases via an increased release of ATP through hemichannels. Of note, an indirect effect of these Cx30 mutants on ATP release by alternative sources, such as pannexin channels (Pnx1) or purinergic receptors (P2XR), cannot be excluded. Besides aberrant hemichannel function, another putative mechanism of disease has also been proposed. The Cx26 mutation associated with Vohwinkel syndrome (D66H) and the GJB2 R75W missense mutation, which is associated with sensori-neural deafness and PPK, are both located in EL1, and both have a dominant negative effect on Cx26 gap junction assembly. It is thus believed that alterations in this domain could impair proper docking of connexons. More recently, an implication of ER stress and subsequent unfolded protein response in the pathogenesis of EKV has been suggested for mutations in Cx31. Three Cx31mutants (G12D, R42P, and C86S) which have an impaired trafficking to the membrane when expressed in transfected cells are associated with decreased cell viability. However, in contrast to Cx26 and Cx30 mutants, cell death induced by these mutants could not be entirely explained by an aberrant opening of hemichannels, as an increase in extracellular Ca^{2+} could not rescue viability. As these mutants were associated with an upregulation of proteasome markers, ER resident proteins, and chaperones, it seems likely that accumulation of EKV-associated Cx31 mutants leads to ER stress and, ultimately, to cell death. However, the mechanism linking increased cell death in vitro to a hyperproliferative skin disorder in vivo remains to be established. Finally, mutations of some specific residues in the NT of connexins are con-sistently found in skin disorders. The Glycine at position 11 or 12 of b-group connexins seems to be of particular importance. The replacement of this glycine by a polar residue has been associated with skin disease in each of these connexins (Cx26 G12R in relation to KID, Cx31 G12R, G12D and Cx30Æ3 G12D in EKV, and Cx30 G11R in Clouston syndrome). It is hypothesized that this conserved glycine maintains the flexibility of the NT and thus enables the gating of the channel by this domain.

12.5 Cx43, Cx37, Cx40, and Cardiovascular Disease

Gap junctions in cardiomyocytes are mainly present at the intercalated disk and are composed of Cx43 in the ventricle and of a combination of Cx40 and Cx43 in the atrium. All three cardiovascular connexins (Cx37, Cx40, and Cx43) are expressed in the vascular wall, with varying expression patterns depending on the localization in the vascular tree. A striking aspect of Cx43 mutations in humans is the lack of cardiac phenotype. Although Cx43 is expressed throughout the heart and is known to play a critical role in cardiac development and function during adult life, a minimal number of ODDD patients actually have cardiac manifestations. A familial case of ODDD with recurrent ventricular tachycardia and atrioventricular block has been associated with the I130T mutation. Interestingly, mice bearing this mutation in a heterozygous state have a phenotype resembling ODDD and display an increased susceptibility to ventricular tachycardia. However, the functional effects of this mutation at the cellular level, that is, a reduced formation of gap junction plaques and a lower conductance between cells, do not distinguish it from other mutations underlying ODDD without any cardiac involvement. As this mutation has only been detected in one family, the concomitant role of other genetic alterations cannot be excluded. One study on six patients with severe forms of visceroatrial heterotaxia, including asplenia or polysplenia and pulmonary atresia or stenosis, detected one or two Cx43 mutations in each of them (all in the CT). Recently, a report on 418 Chinese patients with congenital heart defects described the discovery of three Cx43 mutations in their patients (two in the CT and one in TM3). Thus, the relative contribution of Cx43 mutations to cardiac malformations in humans remains to be clarified. A recent report described Cx40 mutations in patients with idiopathic atrial fibrillation. As this connexin is highly expressed in atrial cardiomyocytes, mutations in this gene are expected to affect atrial conduction properties. In this series of 15 patients with idiopathic atrial fibrillation, four patients displayed mutations in the Cx40 gene, all in the transmembrane domain: two patients exhibited the same P88S heterozygous missense mutation, one patient had a A96S substitution on one allele, and one patient had two different missense mutations (G38D and M163V, the latter being presumably a polymorphism). When examined functionally in transfected cells, these mutations appeared to affect gap junctional communication, either by a defect in localization (G38D, P88S) or by massively reducing gap junctional conductance (P88S, A96S). Interestingly, only the A96S mutation could be detected in peripheral blood lymphocyte DNA, indicating a germ line mutation. The other three patients only had GJA5 mutations in cardiac tissue samples and not in peripheral blood samples, suggesting somatic mutations. The third cardiovascular connexin, Cx37, is mainly expressed in endothelial cells. Interestingly, a Cx37 C1019T genetic polymorphism has been strongly associated with atherosclerosis development and myocardial infarction in humans. Moreover, the resulting amino acid change in the Cx37-CT appeared to alter ATP release through Cx37 hemichannels thereby affecting monocyte adhesion.

Suggested Reading

Abrams CK, Freidin MM, Verselis VK, Bennett MV, Bargiello TA (2001) Functional alterations in gap junction channels formed by mutant forms of connexin 32: evidence for loss of function as a pathogenic mechanism in the X-linked form of Charcot-Marie-Tooth disease. Brain Res 900:9–25

Ahmad S, Evans WH (2002) Post-translational integration and oligomerizationof connexin 26 in plasma membranes and evidence of formation of membrane pores: implications for the assembly of gap junctions. Biochem J 365:693–699

Ahmad S, Chen S, Sun J, Lin X (2003) Connexins 26 and 30 are co-assembled to form gap junctions in the cochlea of mice. Biochem Biophys Res Commun 307:362–368

Ahmad Waza A, Andrabi K, Hussain MU (2012) Adenosine-triphosphate-sensitive K+ channel (Kir6.1): a novel phosphospecific interaction partner of connexin 43 (Cx43). Exp Cell Res 318:2559–2566. doi:10.1016/j.yexcr.2012.08.004

Ai X, Pogwizd SM (2005) Connexin 43 downregulation and dephosphorylation in nonischemic heart failure is associated with enhanced colocalized protein phosphatase type 2A. Circ Res 96:54–63

Ai Z, Fischer A, Spray DC, Brown AM, Fishman GI (2000) Wnt-1 regulation of connexin43 in cardiac myocytes. J Clin Invest 105:161–171

Akazawa H, Komuro I (2003) Too much Csx/Nkx2.5 is as bad as too little? J Mol Cell Cardiol 35:227–229

Anderson C, Catoe H, Werner R (2006) MIR-206 regulates connexin43 expression during skeletal muscle development. Nucleic Acids Res 34:5863–5871

Arita K, Akiyama M, Aizawa T, Umetsu Y, Segawa I, Goto M, Sawamura D, Demura M, Kawano K, Shimizu H (2006) A novel N14Y mutation in Connexin26 in keratitisichthyosis deafness syndrome: analyses of altered gap junctional communication and molecular structure of N terminus of mutated Connexin26. Am J Pathol 169:416–423

Bai S, Spray DC, Burk RD (1993) Identification of proximal and distal regulatory elements of the rat connexin32 gene. Biochim Biophys Acta 1216:197–204

Bai S, Schoenfeld A, Pietrangelo A, Burk RD (1995) Basal promoter of the rat connexin 32 gene: identification and characterization of an essential element and its DNA-binding protein. Mol Cell Biol 15:1439–1445

Bakirtzis G, Choudhry R, Aasen T, Shore L, Brown K, Bryson S, Forrow S, Tetley L, Finbow M, Greenhalgh D, Hodgins M (2003) Targeted epidermal expression of mutant Connexin 26(D66H) mimics true Vohwinkel syndrome and provides a model for the pathogenesis of dominant connexin disorders. Hum Mol Genet 12:1737–1744

Bao X, Lee SC, Reuss L, Altenberg GA (2007) Change in permeant size selectivity by phosphorylation of connexin 43 gap-junctional hemichannels by PKC. Proc Natl Acad Sci U S A 104:4919–4924

© Springer India 2014

M.U. Hussain, *Connexins: The Gap Junction Proteins*, SpringerBriefs in Biochemistry and Molecular Biology, DOI 10.1007/978-81-322-1919-4

Baylin SB (2005) DNA methylation and gene silencing in cancer. Nat Clin Pract Oncol 2:S4–S11

Bazzoni G, Dejana E (2004) Endothelial cell-to-cell junctions: molecular organization and role in vascular homeostasis. Physiol Rev 84:869–901

Beltramello M, Piazza V, Bukauskas FF, Pozzan T, Mammano F (2005) Impaired permeability to Ins(1,4,5)P3 in a mutant connexin underlies recessive hereditary deafness. Nat Cell Biol 7:63–69

Bennett MVL, Goodenough DA (1978) Gap junctions, electronic coupling, and intercellular communication. Neurosci Res Prog Bull 16:373–486

Berthoud VM, Minogue PJ, Laing JG, Beyer EC (2004) Pathways for degradation of connexins and gap junctions. Cardiovasc Res 62:256–267

Bertram JS, Vine AL (2005) Cancer prevention by retinoids and carotenoids: independent action on a common target. Biochim Biophys Acta 1740:170–178

Berube C, Boucher LM, Ma W, Wakeham A, Salmena L, Hakem R, Yeh WC, Mak TW, Benchimol S (2005) Apoptosis caused by p53-induced protein with death domain (PIDD) depends on the death adapter protein RAIDD. Proc Natl Acad Sci U S A 102:4314–14320

Beyer EC, Paul DL, Goodenough DA (1987) Connexin43: a protein from rat heart homologous to a gap junction protein from liver. J Cell Biol 105:2621–2629

Bicego M, Beltramello M, Melchionda S, Carella M, Piazza V, Zelante L, Bukauskas FF, Arslan E, Cama E, Pantano S, Bruzzone R, D'Andrea P, Mammano F (2006) Pathogenetic role of the deafness-related M34T mutation of Cx26. Hum Mol Genet 15:2569–2587

Bicknell KA, Coxon CH, Brooks G (2004) Forced expression of the cyclin B1B1–CDC2 complex induces proliferation in adult rat cardiomyocytes. Biochem J 382:411–416

Boengler K, Dodoni G, Rodriguez-Sinovas A, Cabestrero A, Ruiz-Meana M, Gres P, Konietzka I, Lopez-Iglesias C, Garcia-Dorado D, Di Lisa F, Heusch G, Schulz R (2005) Connexin43 in cardiomyocyte mitochondria and its increase by ischemic preconditioning. Cardiovasc Res 67:234–244

Boerma M, Forsberg L, Van Zeijl L, Morgenstern R, De Faire U, Lemne C et al (1999) A genetic polymorphism in connexin 37 as a prognostic marker for atherosclerotic plaque development. J Intern Med 246:211–218

Borke JL, Yu JC, Isales CM, Wagle N, Do NN, Chen JR, Bollag RJ (2003) Tension-induced reduction in connexin 43 expression in cranial sutures is linked to transcriptional regulation by TBX2. Ann Plast Surg 51:499–504

Bouvier D, Spagnol G, Chenavas S, Kieken F, Vitrac H, Brownell S, Kellezi A, Forge V, Sorgen PL (2009) Characterization of the structure and intermolecular interactions between the connexin40 and connexin43 carboxyl-terminal and cytoplasmic loop domains. J Biol Chem 284:34257–34271

Brigstock DR (2003) The CCN family: a new stimulus package. J Endocrinol 178:169–175

Bruzzone R, Gomes D, Denoyelle E, Duval N, Perea J, Veronesi V, Weil D, Petit C, Gabellec MM, D'Andrea P, White TW (2001) Functional analysis of a dominant mutation of human connexin26 associated with nonsyndromic deafness. Cell Commun Adhes 8:425–431

Burt JM, Steele TD (2003) Selective effect of PDGF on connexin43 versus connexin40 comprised gap junction channels. Cell Commun Adhes 10:287–291

Butkevich E, Hulsmann S, Wenzel D, Shirao T, Duden R, Majoul I (2004) Drebrin is a novel connexin-43 binding partner that links gap junctions to the submembrane cytoskeleton. Curr Biol 14:650–658

Carystinos GD, Kandouz M, Alaoui-Jamali MA, Batist G (2003) Unexpected induction of the human connexin 43 promoter by the ras signaling pathway is mediated by a novel putative promoter sequence. Mol Pharmacol 63:821–831

Cavallaro U, Liebner S, Dejana E (2006) Endothelial cadherins and tumor angiogenesis. Exp Cell Res 312:659–667

Chadjichristos CE, Matter CM, Roth I, Sutter E, Pelli G, Luscher TF, Chanson M, Kwak BR (2006) Reduced connexin43 expression limits neointima formation after balloon distension injury in hypercholesterolemic mice. Circulation 113:2835–2843

Chen ZQ, Lefebvre D, Bai XH, Reaume A, Rossant J, Lye SJ (1995) Identification of two regulatory elements within the promoter region of the mouse connexin 43 gene. J Biol Chem 270:3863–3868

Chen Y, Hühn D, Knösel T, Pacyna-Gengelbach M, Deutschmann N, Petersen I (2005) Downregulation of connexin 26 in human lung cancer is related to promoter methylation. Int J Cancer 113:14–21

Cheng A, Tang H, Cai J, Zhu M, Zhang X, Rao M, Mattson MP (2004) Gap junctional communication is required to maintain mouse cortical neural progenitor cells in a proliferative state. Dev Biol 272:203–216

Cherian PP, Siller-Jackson AJ, Gu S, Wang X, Bonewald LF, Sprague E, Jiang JX (2005) Mechanical strain opens connexin 43 hemichannels in osteocytes: a novel mechanism for the release of prostaglandin. Mol Biol Cell 16:3100–3106

Chesik D, Glazenburg K, Wilczak N, Geeraedts F, De Keyser J (2004) Insulin-like growth factor binding protein-1-6 expression in activated microglia. Neuroreport 15:1033–1037

Chu CY, Rana TM (2007) Small RNAs: regulators and guardians of the genome. J Cell Physiol 213:412–419

Civitelli R, Ziambaras K, Warlow PM, Lecanda F, Nelson T, Harley J, Atal N, Beyer EC, Steinberg TH (1998) Regulation of connexin43 expression and function by prostaglandin E2 (PGE2) and parathyroid hormone (PTH) in osteoblastic cells. J Cell Biochem 68:8–21

Cohen-Salmon M, Regnault B, Cayet N, Caille D, Demuth K, Hardelin JP, Janel N, Meda P, Petit C (2007) Connexin30 deficiency causes intrastrial fluid-blood barrier disruption within the cochlear stria vascularis. Proc Natl Acad Sci U S A 104:6229–6234

Common JE, Becker D, Di WL, Leigh IM, O'Toole EA, Kelsell DP (2002) Functional studies of human skin disease- and deafness-associated connexin 30 mutations. Biochem Biophys Res Commun 298:651–656

Condorelli DF, Parenti R, Spinella F, Trovato Salinaro A, Belluardo N, Cardile V, Cicirata F (1998) Cloning of a new gap junction gene (Cx36) highly expressed in mammalian brain neurons. Eur J Neurosci 10:1202–1208

Cottrell GT, Burt JM (2001) Heterotypic gap junction channel formation between heteromeric and homomeric Cx40 and Cx43connexons. Am J Physiol Cell Physiol 281:C1559–C1567

Cruciani V, Mikalsen SO (2006) The vertebrate connexin family. Cell Mol Life Sci 63:1125–1140

Cruciani V, Mikalsen SO (2007) Evolutionary selection pressure and family relationships among connexin genes. Biol Chem 388:253–264

Dagli ML, Yamasaki H, Krutovskikh V, Omori Y (2004) Delayed liver regeneration and increased susceptibility to chemical hepatocarcinogenesis in transgenic mice expressing a dominant-negative mutant of connexin32 only in the liver. Carcinogenesis 25:483–492

Dahl G, Locovei S (2006) Pannexin to gap or not to gap, is that a question? IUBMB Life 58:409–419

Dahl G, Werner R, Levine E, Rabadan-Diehl C (1992) Mutational analysis of gap junction formation. Biophys J 62:172–180

D'Ambrosio R, Wenzel J, Schwartzkroin PA, McKhann GM, Janigro D II (1998) Functional specialization and topographic segregation of hippocampal astrocytes. J Neurosci 18:4425–4438

Dang X, Doble BW, Kardami E (2003) The carboxy-tail of connexin-43 localizes to the nucleus and inhibits cell growth. Mol Cell Biochem 242:35–38

Dang X, Jeyaraman M, Kardami E (2006) Regulation of connexin-43-mediated growth inhibition by a phosphorylatable amino-acid is independent of gap junction-forming ability. Mol Cell Biochem 289:201–207

Deans MR, Gibson JR, Sellitto C, Connors BW, Paul DL (2001) Synchronous activity of inhibitory networks in neocortex requires electrical synapses containing connexin36. Neuron 31:477–485

Debeer P, Van Esch H, Huysmans C, Pijkels E, De Smet L, Van de Ven W et al (2005) Novel GJA1 mutations in patients with oculo-dento-digital dysplasia (ODDD). Eur J Med Genet 48:377–387

Debrus S, Tuffery S, Matsuoka R, Galal O, Sarda P, Sauer U et al (1997) Lack of evidence for connexin 43 gene mutations in human autosomal recessive lateralization defects. J Mol Cell Cardiol 29:1423–1431

De Leon JR, Buttrick PM, Fishman GI (1994) Functional analysis of the connexin43 gene promoter in vivo and in vitro. J Mol Cell Cardiol 26(3):379–389

Delmar M, Coombs W, Sorgen P, Duffy HS, Taffet SM (2004) Structural bases for the chemical regulation of Connexin43 channels. Cardiovasc Res 62:268

Deng Y, Chen Y, Reuss L, Altenberg GA (2006) Mutations of connexin 26 at position 75 and dominant deafness: essential role of arginine for the generation of functional gapjunctional channels. Hear Res 220:87–94

Dermietzel R (1996) Molecular diversity and plasticity of gap junctions in the nervous system. In: Spray DC, Dermietzel R (eds) Gap junctions in the nervous system. Chapman & Hall, New York, pp 13–38

Dermietzel R, Spray DC (1993) Gap junctions in the brain: where, what type, how many and why? Trends Neurosci 16:186–192

Dermietzel R, Hertberg EL, Kessler JA, Spray DC (1991) Gap junctions between cultured astrocytes: immunocytochemical, molecular, and electrophysiological analysis. J Neurosci 11:1421–1432

Dermietzel R, Farooq M, Kessler JA, Althaus H, Hertzberg EL, Spray DC (1997) Oligodendrocytes express gap junction proteins connexin32 and connexin45. Glia 20:101–114

Dermietzel R, Gao Y, Scemes E, Vieira D, Urban M, Kremer M, Bennett MV, Spray DC (2000a) Connexin43 null mice reveal that astrocytes express multiple connexins. Brain Res Brain Res Rev 32:45–56

Dermietzel R, Kremer M, Paputsoglu G, Stang A, Skerrett IM, Gomes D, Srinivas M, Janssen-Bienhold U, Weiler R, Nicholson BJ, Bruzzone R, Spray DC (2000b) Molecular and functional diversity of neural connexins in the retina. J Neurosci 20:8331–8343

DeRosa AM, Xia CH, Gong X, White TW (2007) The cataract-inducing S50P mutation in Cx50 dominantly alters the channel gating of wild-type lens connexins. J Cell Sci 120:4107–4116

Derouette JP, Wong C, Burnier L, Morel S, Sutter E, Galan K et al (2009) Molecular role of Cx37 in advanced atherosclerosis: a micro-array study. Atherosclerosis 206:69–76

Dhein S (2005) New, emerging roles for cardiac connexins: mitochondrial Cx43 raises new questions. Cardiovasc Res 67:179–181

Di WL, Monypenny J, Common JE, Kennedy CT, Holland KA, Leigh IM, Rugg EL, Zicha D, Kelsell DP (2005) Defective trafficking and cell death is characteristic of skin disease associated connexin 31 mutations. Hum Mol Genet 11:2014–2022

Dirnagl U, Iadecola C, Moskowitz MA (1999) Pathobiology of ischaemic stroke: an integrated view. Trends Neurosci 22:391–397

Doble BW, Kardami E (1995) Basic fibroblast growth factor stimulates connexin-43 expression and intercellular communication of cardiac fibroblasts. Mol Cell Biochem 143:81–87

Doble BW, Woodgett JR (2003) GSK-3: tricks of the trade for a multi-tasking kinase. J Cell Sci 116:1175–1186

Doble BW, Chen Y, Bosc DG, Litchfield DW, Kardami E (1996) Fibroblast growth factor-2 decreases metabolic coupling and stimulates phosphorylation as well as masking of connexin43 epitopes in cardiac myocytes. Circ Res 79:647–658

Doble BW, Ping P, Kardami E (2000) The epsilon subtype of protein kinase C is required for cardiomyocyte connexin-43 phosphorylation. Circ Res 86:293–301

Doble BW, Dang X, Ping P, Fandrich RR, Nickel BE, Jin Y, Cattini PA, Kardami E (2004) Phosphorylation of serine 262 in the gap junction protein connexin-43 regulates DNA synthesis in cell–cell contact forming cardiomyocytes. J Cell Sci 117:507–514

Dobrowolski R, Sommershof A, Willecke K (2007) Some oculodentodigital dysplasia-associated Cx43 mutations cause increased hemichannel activity in addition to deficient gap junction channels. J Membr Biol 219:9–17

Draguhn A, Traub RD, Schmitz D, Jefferys JG (1998) Electrical coupling underlies high-frequency oscillations in the hippocampus in vitro. Nature 394:189–192

Duffy HS, Delmar M, Spray DC (2002) Formation of the gap junction nexus: binding partners for connexins. J Physiol Paris 96:243–249

Echetebu CO, Ali M, Izban MG, MacKay L, Garfield RE (1999) Localization of regulatory protein binding sites in the proximal region of human myometrial connexin 43 gene. Mol Hum Reprod 5:757–766

Eiberger J, Degen J, Romualdi A, Deutsch U, Willecke K, Sohl G (2001) Connexin genes in the mouse and human genome. Cell Commun Adhes 8:163–165

Elisevich K, Rempel SA, Smith B, Allar N (1997) Connexin43 mRNA expression in two experimental models of epilepsy. Mol Chem Neuropathol 32:75–88

el-Sabban ME, Pauli BU (1991) Cytoplasmic dye transfer between metastatic tumor cells and vascular endothelium. J Cell Biol 115:1375–1382

El-Sabban ME, Sfeir AJ, Daher MH, Kalaany NY, Bassam RA, Talhouk RS (2003) ECM-induced gap junctional communication enhances mammary epithelial cell differentiation. J Cell Sci 116:3531–3541

Evans WH, Martin PE (2002) Gap junctions: structure and function (review). Mol Membr Biol 19:121–136

Evans WH, De Vuyst E, Leybaert L (2006) The gap junction cellular internet: connexin hemichannels enter the signalling limelight. Biochem J 397:1–14

Evert M, Ott T, Temme A, Willecke K, Dombrowski F (2002) Morphology and morphometric investigation of hepatocellular preneoplastic lesions and neoplasms in connexin32-deficient mice. Carcinogenesis 23:697–703

Fernandez-Cobo M, Gingalewski C, De Maio A (1998) Expression of the connexin 43 gene is increased in the kidneys and the lungs of rats injected with bacterial lipopolysaccharide. Shock 10:97–102

Filipowicz W, Bhattacharyya SN, Sonenberg N (2008) Mechanisms of post-transcriptional regulation by microRNAs: are the answers in sight. Nat Rev Genet 9:102–114

Fletcher WH, Byus CV, Walsh DA (1987) Receptor-mediated action without receptor occupancy: a function for cell–cell communication in ovarian follicles. Adv Exp Med Biol 219:299–323

Franco L, Zocchi E, Usai C, Guida L, Bruzzone S, Costa A, De Flora A (2001) Paracrine roles of NAD+ and cyclic ADP-ribose in increasing intracellular calcium and enhancing cell proliferation of 3T3 fibroblasts. J Biol Chem 276:21642–21648

Fu CT, Bechberger JF, Ozog MA, Perbal B, Naus CC (2004) CCN3 (NOV) interacts with connexin43 in C6 glioma cells: possible mechanism of connexin-mediated growth suppression. J Biol Chem 279:36943–36950

Gabriel HD, Strobl B, Hellmann P, Buettner R, Winterhager E (2001) Organization and regulation of the rat Cx31 gene. Implication for a crucial role of the intron region. Eur J Biochem 268:1749–1759

Gangenahalli GU, Singh VK, Verma YK, Gupta P, Sharma RK, Chandra R, Gulati S, Luthra PM (2005) Three-dimensional structure prediction of the interaction of CD34 with the SH3 domain of Crk-L. Stem Cells Dev 14:470–477

Geimonen E, Jiang W, Ali M, Fishman GI, Garfield RE, Andersen J (1996) Activation of protein kinase C in human uterine smooth muscle induces connexin-43 gene transcription through an AP-1 site in the promoter sequence. J Biol Chem 271:23667–23674

Gellhaus A, Dong X, Propson S, Maass K, Klein-Hitpass L, Kibschull M, Traub O, Willecke K, Perbal B, Lye SJ, Winterhager E (2004) Connexin43 interacts with NOV: a possible mechanism for negative regulation of cell growth in choriocarcinoma cells. J Biol Chem 279:36931–36942

George CH, Kendall JM, Evans WH (1999) Intracellular trafficking pathways in the assembly of connexins into gap junctions. J Biol Chem 274:8678–8685

Gerido DA, White TW (2004) Connexin disorders of the ear, skin, and lens. Biochim Biophys Acta 1662:159–170

Giebel J, Woenckhaus C, Fabian M, Tost F (2005) Age-related differential expression of apoptosis-related genes in conjunctival epithelial cells. Acta Ophthalmol Scand 83:471–476

Giepmans BN (2004) Gap junctions and connexin-interacting proteins. Cardiovasc Res 62:233–245

Giepmans BN (2006) Role of connexin43-interacting proteins at gap junctions. Adv Cardiol 42:41–56

Giepmans BN, Moolenaar WH (1998) The gap junction protein connexin43 interacts with the second PDZ domain of the zona occludens-1 protein. Curr Biol 8:931–934

Giepmans BN, Feiken E, Gebbink MF, Moolenaar WH (2003) Association of connexin43 with a receptor protein tyrosine phosphatase. Cell Commun Adhes 10:201–205

Go M, Kojima T, Takano K, Murata M, Koizumi J, Kurose M, Kamekura R, Osanai M, Chiba H, Spray DC, Himi T, Sawada N (2006) Connexin26 expression prevents down-regulation of barrier and fence functions of tight junctions by Na+/K+-ATPase inhibitor ouabain in human airway epithelial cell line calu-3. Exp Cell Res 312:3847–3856

Goldberg GS, Lampe PD, Nicholson BJ (1999) Selective transfer of endogenous metabolites through gap junctions composed of different connexins. Nat Cell Biol 1:457–459

Goldberg GS, Bechberger JF, Tajima Y, Merritt M, Omori Y, Gawinowicz MA, Narayanan R, Tan Y, Sanai Y, Yamasaki H, Naus CC, Tsuda H, Nicholson BJ (2000) Connexin43 suppresses MFG-E8 while inducing contact growth inhibition of glioma cells. Cancer Res 60:6018–6026

Goldberg GS, Moreno AP, Lampe PD (2002) Gap junctions between cells expressing connexin43 or 32 show inverse perm selectivity to adenosine and ATP. J Biol Chem 277:36725–36730

Gollob MH, Jones DL, Krahn AD, Danis L, Gong XQ, Shao Q, Liu X, Veinot JP, Tang AS, Stewart AF, Tesson F, Klein GJ, Yee R, Skanes AC, Guiraudon GM, Ebihara L, Bai D (2006) Somatic mutations in the connexin40 gene (GJA5) in atrial fibrillation. N Engl J Med 354:2677–2688

Gong X, Li E, Klier G, Huang Q, Wu Y, Lei H et al (1997) Disruption of alpha3 connexin gene leads to proteolysis and cataractogenesis in mice. Cell 91:833–843

Gonzalez-Mariscal L, Nava P (2005) Tight junctions, from tight intercellular seals to sophisticated protein complexes involved in drug delivery, pathogens interaction and cell proliferation. Adv Drug Deliv Rev 57:811–814

Gonzalez-Mariscal L, Betanzos A, Nava P, Jaramillo BE (2003) Tight junction proteins. Prog Biophys Mol Biol 81:1–44

Goodenough DA, Paul DL (2003) Beyond the gap: functions of unpaired connexon channels. Nat Rev Mol Cell Biol 4:285–294

Gottardi CJ, Gumbiner BM (2001) Adhesion signaling: how beta-catenin interacts with its partners. Curr Biol 11:R792–R794

Govindarajan R, Zhao S, Song XH, Guo RJ, Wheelock M, Johnson KR, Mehta PP (2002) Impaired trafficking of connexins in androgen-independent human prostate cancer cell lines and its mitigation by alpha-catenin. J Biol Chem 277:50087–50097

Graeber SH, Hulser DF (1998) Connexin transfection induces invasive properties in HeLa cells. Exp Cell Res 243:142–149

Grifa A, Wagner CA, D'Ambrosio L, Melchionda S, Bernardi F, Lopez-Bigas N, Rabionet R, Arbones M, Monica MD, Estivill X, Zelante L, Lang F, Gasparini P (1999) Mutations in GJB6 cause nonsyndromic autosomal dominant deafness at DFNA3 locus. Nat Genet 23:16–18

Groenewegen WA, van Veen TA, van der Velden HM, Jongsma HJ (1998) Genomic organization of the rat connexin40 gene: identical transcription start sites in heart and lung. Cardiovasc Res 38:463–471

Guimond J, Devost D, Brodeur H, Mader S, Bhat PV (2002) Characterization of the rat RALDH1 promoter. A functional CCAAT and octamer motif are critical for basal promoter activity. Biochim Biophys Acta 1579:81–91

Gutstein DE, Morley GE, Tamaddon H, Vaidya D, Schneider MD, Chen J, Chien KR, Stuhlmann H, Fishman GI (2001) Conduction slowing and sudden arrhythmic death in mice with cardiac-restricted inactivation of connexin43. Circ Res 88:333–339

Hagendorff A, Schumacher B, Kirchhoff S, Luderitz B, Willecke K (1999) Conduction disturbances and increased atrial vulnerability in Connexin40-deficient mice analyzed by transesophageal stimulation. Circulation 99:1508–1515

Hansen L, Yao W, Eiberg H, Funding M, Riise R, Kjaer KW et al (2006) The congenital "ant-egg" cataract phenotype is caused by a missense mutation in connexin46. Mol Vis 12:1033–1039

Harris AL (2001) Emerging issues of connexin channels: biophysics fills the gap. Q Rev Biophys 34:325–472

Hattori Y, Fukushima M, Maitani Y (2007) Non-viral delivery of the connexin 43 gene with histone deacetylase inhibitor to human nasopharyngeal tumor cells enhances gene expression and inhibits in vivo tumor growth. Int J Oncol 30:1427–1439

He DS, Jiang JX, Taffet SM, Burt JM (1999) Formation of heteromeric gap junction channels by connexin 40 and 43 in vascular smooth muscle cells. Proc Natl Acad Sci U S A 96:6495–6500

Hennemann H, Kozjek G, Dahl E, Nicholson B, Willecke K (1992) Molecular cloning of mouse connexins26 and -32: similar genomic organization but distinct promoter sequences of two gap junction genes. Eur J Cell Biol 58:81–89

Herve JC, Sarrouilhe D (2006) Protein phosphatase modulation of the intercellular junctional communication: importance in cardiac myocytes. Prog Biophys Mol Biol 90:225–248

Herve JC, Bourmeyster N, Sarrouilhe D (2004) Diversity in protein–protein interactions of connexins: emerging roles. Biochim Biophys Acta 1662:22–41

Hibino H, Kurachi Y (2006) Molecular and physiological bases of the K+ circulation in the mammalian inner ear. Physiology (Bethesda) 21:336–345

Hirai A, Yano T, Nishikawa K, Suzuki K, Asano R, Satoh H, Hagiwara K, Yamasaki H (2003) Down-regulation of connexin 32 gene expression through DNA methylation in a human renal cell carcinoma cell. Am J Nephrol 23:172–177

Holcik M, Korneluk RG (2000) Functional characterization of the X-linked inhibitor of apoptosis (XIAP) internal ribosome entry site element: role of La autoantigen in XIAP translation. Mol Cell Biol 20:4648–4657

Holnthoner W, Pillinger M, Groger M, Wolff K, Ashton AW, Albanese C, Neumeister P, Pestell RG, Petzelbauer P (2002) Fibroblast growth factor-2 induces Lef/Tcf-dependent transcription in human endothelial cells. J Biol Chem 277:45847–45853

Hormuzdi SG, Pais I, LeBeau FE, Towers SK, Rozov A, Buhl EH, Whittington MA, Monyer H (2001) Impaired electrical signaling disrupts gamma frequency oscillations in connexin 36-deficient mice. Neuron 31:487–495

Huang R, Lin Y, Wang CC, Gano J, Lin B, Shi Q, Boynton A, Burke J, Huang RP (2002) Connexin43 suppresses human glioblastoma cell growth by down-regulation of monocyte chemotactic protein 1, as discovered using protein array technology. Cancer Res 62:2806–2812

Hudder A, Werner R (2000) Analysis of a Charcot-Marie-Tooth disease mutation reveals an essential internal ribosome entry site element in the connexin-32 gene. J Biol Chem 275:34586–34591

Hunter AW, Barker RJ, Zhu C, Gourdie RG (2005) Zonula occludens-1 alters connexin43 gap junction size and organization by influencing channel accretion. Mol Biol Cell 16:5686–5698

Hussain MU, Zoidl G, Klooster J, Kamermans M, Dermietzel R (2008) IRES-mediated translation of the carboxy-terminal domain of the horizontal cell specific connexin Cx55.5 in vivo and in vitro. BMC Mol Biol 9:52

Iacobas DA, Urban-Maldonado M, Iacobas S, Scemes E, Spray DC (2003a) Array analysis of gene expression in connexin-43 null astrocytes. Physiol Genomics 15:177–190

Iacobas DA, Urban M, Iacobas S, Spray DC (2003b) Transcription regulation and coordination of some cell signaling genes in brain and heart of connexin43 null mouse. Rev Med Chir Soc Med Nat Iasi 107:534–539

Iacobas DA, Urban M, Iacobas S, Spray DC (2003c) Transcriptomic characterization of four classes of cell–cell/matrix genes in brains and hearts of wild type and connexin43 null mice. Rom J Physiol 40:68–85

Iacobas DA, Scemes E, Spray DC (2004) Gene expression alterations in connexin null mice extend beyond the gap junction. Neurochem Int 45:243–250

Iacobas DA, Iacobas S, Li WE, Zoidl G, Dermietzel R, Spray DC (2005a) Genes controlling multiple functional pathways are transcriptionally regulated in connexin43 null mouse heart. Physiol Genomics 20:211–223

Iacobas DA, Iacobas S, Urban-Maldonado M, Spray DC (2005b) Sensitivity of the brain transcriptome to connexin ablation. Biochim Biophys Acta 1711:183–196

Iacobas DA, Iacobas S, Spray DC (2007) Connexin43 and the brain transcriptome of the newborn mice. Genomics 89:113–123

Israsena N, Hu M, Fu W, Kan L, Kessler JA (2004) The presence of FGF2 signaling determines whether beta-catenin exerts effects on proliferation or neuronal differentiation of neural stem cells. Dev Biol 268:220–231

Jacob A, Beyer EC (2001) Mouse connexin 45: genomic cloning and exon usage. DNA Cell Biol 20:11–19

Jacque E, Tchenio T, Piton G, Romeo PH, Baud V (2005) RelA repression of RelB activity induces selective gene activation downstream of TNF receptors. Proc Natl Acad Sci U S A 102:14635–14640

Jandt E, Denner K, Kovalenko M, Ostman A, Bohmer FD (2003) The protein-tyrosine phosphatase DEP-1 modulates growth factor-stimulated cell migration and cell–matrix adhesion. Oncogene 22:4175–4185

Janssens S, Tinel A, Lippens S, Tschopp J (2005) PIDD mediates NF-kappaB activation in response to DNA damage. Cell 123:1079–1092

Jiang JX, Gu S (2005) Gap junction- and hemichannel-independent actions of connexins. Biochim Biophys Acta 1711:208–214

John SA, Revel JP (1991) Connexon integrity is maintained by noncovalent bonds:Intramolecular disulfide bonds link the extracellular domains in rat connexin-43. Biochem Biophys Res Commun 178:1312–1318

Johnson RG, Meyer RA, Li XR, Preus DM, Tan L, Grunenwald H, Paulson AF, Laird DW, Sheridan JD (2002) Gap junctions assemble in the presence of cytoskeletal inhibitors, but enhanced assembly requires microtubules. Exp Cell Res 275:67–80

Jones SA, Lancaster MK, Boyett MR (2004) Ageing-related changes of connexins and conduction within the sinoatrial node. J Physiol 560:429–437

Jongen WM, Fitzgerald DJ, Asamoto M, Piccoli C, Slaga TJ, Gros D, Takeichi M, Yamasaki H (1991) Regulation of connexin 43-mediated gap junctional intercellular communication by Ca2+ in mouse epidermal cells is controlled by E-cadherin. J Cell Biol 114:545–555

Jordan K, Chodock R, Hand AR, Laird DW (2001) The origin of annular junctions: a mechanism of gap junction internalization. J Cell Sci 114:763–773

Judisch GF, Martin-Casals AM, Hanson JW, Olin WH (1979) Oculodendrodigital dysplasia; four new reports and a literature review. Arch Ophtalmol 97:878–884

Kamei J, Toyofuku T, Hori M (2003) Negative regulation of p21 by [beta]-catenin/TCF signaling: a novel mechanism by which cell adhesion molecules regulate cell proliferation. Biochem Biophys Res Commun 312:380

Kamermans M, Fahrenfort I, Schultz K, Janssen-Bienhold U, Sjoerdsma T, Weiler R (2001) Hemichannel-mediated inhibition in the outer retina. Science 292:1178–1180

Kardami E, Banerji S, Doble BW, Dang X, Fandrich RR, Jin Y, Cattini PA (2003) PKC-dependent phosphorylation may regulate the ability of connexin43 to inhibit DNA synthesis. Cell Commun Adhes 10:293–297

Kardami E, Dang X, Iacobas DA, Nickel BE, Jeyaraman M, Srisakuldee W, Makazan J, Tanguy S, Spray DC (2007) The role of connexins in controlling cell growth and gene expression. Prog Biophys Mol Biol 94:245–264

Kasahara H, Wakimoto H, Liu M, Maguire CT, Converso KL, Shioi T, Huang WY, Manning WJ, Paul D, Lawitts J, Berul CI, Izumo S (2001) Progressive atrioventricular conduction defects and heart failure in mice expressing a mutant Csx/Nkx2.5 homeoprotein. J Clin Invest 108:189–201

Kasahara H, Ueyama T, Wakimoto H, Liu MK, Maguire CT, Converso KL, Kang PM, Manning WJ, Lawitts J, Paul DL, Berul CI, Izumo S (2003) Nkx2.5 homeoprotein regulates expression of gap junction protein connexin 43 and sarcomere organization in postnatal cardiomyocytes. J Mol Cell Cardiol 35:243–256

Kaushansky K, Shoemaker SG, O'Rork CA, McCarty JM (1994) Coordinate regulation of multiple human lymphokine genes by Oct-1 and potentially novel 45 and 43 kDa polypeptides. J Immunol 152:1812–1820

Kelsell DP, Dunlop J, Stevens HP, Lench NJ, Liang JN, Parry G, Mueller RF, Leigh IM (1997) Connexin 26 mutations in hereditary non-syndromic sensorineural deafness. Nature 387:80–83

Kiang DT, Jin N, Tu ZJ, Lin HH (1997) Upstream genomic sequence of the human connexin26 gene. Gene 199:165–171

Kibschull M, Magin TM, Traub O, Winterhager E (2005) Cx31 and Cx43 double-deficient mice reveal independent functions in murine placental and skin development. Dev Dyn 233:853–863

King TJ, Bertram JS (2005) Connexins as targets for cancer chemoprevention and chemotherapy. Biochim Biophys Acta 1719:146–160

King TJ, Lampe PD (2004) The gap junction protein connexin32 is a mouse lung tumor suppressor. Cancer Res 64:7191–7196

Kirchhoff S, Nelles E, Hagendorff A, Krüger O, Traub O, Willecke K (1998) Reduced cardiac conduction velocity and predisposition to arrhythmias in connexin40-deficient mice. Curr Biol 8:299–302

Kojima T, Kokai Y, Chiba H, Yamamoto M, Mochizuki Y, Sawada N (2001) Cx32 but not Cx26 is associated with tight junctions in primary cultures of rat hepatocytes. Exp Cell Res 263:193–201

Kouzarides T (2007) Chromatin modifications and their function. Cell 128:693–705

Koval M (2006) Pathways and control of connexin oligomerization. Trends Cell Biol 16:159–166

Koval M, Harley JE, Hick E, Steinberg TH (1997) Connexin46 is retained as monomers in a trans-golgi compartment of osteoblastic cells. J Cell Biol 137:847–857

Kozak M (1989) The scanning model for translation: an update. J Cell Biol 108:229–241

Kumai M, Nishii K, Nakamura K, Takeda N, Suzuki M, Shibata Y (2000) Loss of connexin45 causes a cushion defect in early cardiogenesis. Development 127:3501–3512

Kumar NM, Gilula NB (1996) The gap junction communication channel. Cell 84:381–388

Kunzelmann P, Blumcke I, Traub O, Dermietzel R, Willecke K (1997) Co-expression of connexin45 and -32 in oligodendrocytes of rat brain. J Neurocytol 26:17–22

Kunzelmann P, Schroder W, Traub O, Steinhauser C, Dermietzel R, Willecke K (1999) Late onset and increasing expression of the gap junction protein connexin30 in adult murine brain and long-term culturedastrocytes. Glia 25:111–119

Kwak BR, Mulhaupt F, Veillard N, Gros DB, Mach F (2002) Altered pattern of vascular connexin expression in atherosclerotic plaques. Arterioscler Thromb Vasc Biol 22:225–230

Kwak BR, Veillard N, Pelli G, Mulhaupt F, James RW, Chanson M, Mach F (2003) Reduced connexin43 expression inhibits atherosclerotic lesion formation in low-density lipoprotein receptor-deficient mice. Circulation 107:1033–1039

Laing JG, Manley-Markowski RN, Koval M, Civitelli R, Steinberg TH (2001) Connexin45 interacts with zonula occludens-1 and connexin43 in osteoblastic cells. J Biol Chem 276:23051–23055

Laird DW (2005) Connexin phosphorylation as a regulatory event linked to gap junction internalization and degradation. Biochim Biophys Acta 1711:172–182

Laird DW (2006) Life cycle of connexins in health and disease. Biochem J 394:527–543

Lampe PD, Lau AF (2000) Regulation of gap junctions by phosphorylation of connexins. Arch Biochem Biophys 384:205–215

Lampe PD, Lau AF (2004) The effects of connexin phosphorylation on gap junctional communication. Int J Biochem Cell Biol 36:1171–1186

Lampugnani MG, Zanetti A, Corada M, Takahashi T, Balconi G, Breviario F, Orsenigo F, Cattelino A, Kemler R, Daniel TO, Dejana E (2003) Contact inhibition of VEGF-induced proliferation requires vascular endothelial cadherin, {beta}-catenin, and the phosphatase DEP-1/CD148. J Cell Biol 161:793–804

Landisman CE, Long MA, Beierlein M, Deans MR, Paul DL, Connors BW (2002) Electrical synapses in the thalamic reticular nucleus. J Neurosci 22:1002–1009

Lasater EM (1987) Retinal horizontal cell gap junctional conductance is modulated by dopamine through a cyclic AMP-dependent protein kinase. Proc Natl Acad Sci U S A 84:7319–7323

Lau AF, Kanemitsu MY, Kurata WE, Danesh S, Boynton AL (1992) Epidermal growth factor disrupts gap-junctional communication and induces phosphorylation of connexin43 on serine. Mol Biol Cell 3:865–874

Le AC, Musil LS (2001) A novel role for FGF and extracellular signal-regulated kinase in gap junction-mediated intercellular communication in the lens. J Cell Biol 154:197–216

Lecanda F, Warlow PM, Sheikh S, Furlan F, Steinberg TH, Civitelli R (2000) Connexin43 deficiency causes delayed ossification, craniofacial abnormalities, and osteoblast dysfunction. J Cell Biol 151:931–944

Lee SW, Tomasetto C, Paul D, Keyomarsi K, Sager R (1992) Transcriptional downregulation of gap-junction proteins blocks junctional communication in human mammary tumor cell lines. J Cell Biol 118:1213–1221

Lee JR, DeRosa AM, White TW (2009) Connexin mutations causing skin disease and deafness increase hemichannel activity and cell death when expressed in Xenopus oocytes. J Invest Dermatol 129:870–878

Levine AJ, Brivanlou AH (2006) GDF3, a BMP inhibitor, regulates cell fate in stem cells and early embryos. Development 133:209–216

Li WE, Nagy JI (2000) Connexin43 phosphorylation state and intercellular communication in cultured astrocytes following hypoxia and protein phosphatase inhibition. Eur J Neurosci 12:2644–2650

Li J, Hertzberg EL, Nagy JI (1997) Connexin32 in oligodendrocytes and association with myelinated fibers in mouse and rat brain. J Comp Neurol 379:571–591

Li J, Shen H, Naus CC, Zhang L, Carlen PL (2001) Upregulation of gap junction connexin32 with epileptiform activity in the isolated mouse hippocampus. Neuroscience 105:589–598

Li L, Guris DL, Okura M, Imamoto A (2003) Translocation of CrkL to focal adhesions mediates integrin-induced migration downstream of Src family kinases. Mol Cell Biol 23:2883–2892

Liao Y, Day KH, Damon DN, Duling BR (2001) Endothelial cell-specific knockout of connexin43 causes hypotension and bradycardia in mice. Proc Natl Acad Sci U S A 98:9989–9994

Lin Y, Ma W, Benchimol S (2000) Pidd, a new death-domain-containing protein, is induced by p53 and promotes apoptosis. Nat Genet 26:122–127

Lin Liang GS, de Miguel M, Gomez-Hernandez JM, Glass JD, Scherer SS, Mintz M et al (2005) Severe neuropathy with leaky connexin32 hemichannels. Ann Neurol 57:749–754

Liu XZ, Xia XJ, Xu LR, Pandya A, Liang CY, Blanton SH, Brown SD, Steel KP, Nance WE (2000) Mutations in connexin31 underlie recessive as well as dominant non-syndromic hearing loss. Hum Mol Genet 9:63–67

Liverman CS, Cui L, Yong C, Choudhuri R, Klein RM, Welch KM, Berman NE (2004) Response of the brain to oligemia: gene expression, c-Fos, and Nrf2 localization. Brain Res Mol Brain Res 126:57–66

Loewenstein WR, Kanno Y (1966) Intercellular communication and the control of tissue growth: lack of communication between cancer cells. Nature 209:1248–1249

Loewenstein WR, Rose B (1992) The cell–cell channel in the control of growth. Semin Cell Biol 3:59–79

Lu C, Zhang DQ, McMahon DG (1999) Electrical coupling of retinal horizontal cells mediated by distinct voltage-independent junctions. Vis Neurosci 16:811–818

Lujambio A, Esteller M (2007) CpG island hypermethylation of tumor suppressor microRNAs in human cancer. Cell Cycle 6:1455–1459

Macari F, Landau M, Cousin P, Mevorah B, Brenner S, Panizzon R, Schorderet DF, Hohl D, Huber M (2000) Mutation in the gene for connexin 30.3 in a family with erythrokeratodermia variabilis. Am J Hum Genet 67:1296–1301

Mackay D, Ionides A, Kibar Z, Rouleau G, Berry V, Moore A, Shiels A, Bhattacharya S (1999) Connexin46 mutations in autosomal dominant congenital cataract. Am J Hum Genet 64:1357–1364

Maestrini E, Korge BP, Ocana-Sierra J, Calzolari E, Cambiaghi S, Scudder PM, Hovnanian A, Monaco AP, Munro CS (1999) A missense mutation in connexin26, D66H, causes mutilating keratoderma with sensorineural deafness (Vohwinkel's syndrome) in three unrelated families. Hum Mol Genet 8:1237–1243

Mann-Metzer P, Yarom Y (1999) Electrotonic coupling interacts with intrinsic properties to generate synchronized activity in cerebellar networks of inhibitory interneurons. J Neurosci 19:3298–3306

Manthey D, Bukauskas F, Lee CG, Kozak CA, Willecke K (1999) Molecular cloning and functional expression of the mouse gap junction gene connexin 57 in human HeLa cells. J Biol Chem 274:14716–14723

Mantz J, Cordier J, Giaume C (1993) Effects of general anesthetics on intercellular communications mediated by gap junctions between astrocytes in primary culture. Anesthesiology 78:892–901

Martin PE, Blundell G, Ahmad S, Errington RJ, Evans WH (2001) Multiple pathways in the trafficking and assembly of connexin 26, 32 and 43 into gap junction intercellular communication channels. J Cell Sci 114:3845–3855

Martinez AD, Saez JC (2000) Regulation of astrocyte gap junctions by hypoxia-reoxygenation. Brain Res Brain Res Rev 32:250–258

Martyn KD, Kurata WE, Warn-Cramer BJ, Burt JM, TenBroek E, Lau AF (1997) Immortalized connexin43 knockout cell lines display a subset of biological properties associated with the transformed phenotype. Cell Growth Differ 8:1015–1027

Matsumoto A, Arai Y, Urano A, Hyodo S (1992) Effect of androgen on the expression of gap junction and beta-actin mRNAs in adult rat motoneurons. Neurosci Res 14:133–144

Meda P (1996) Gap junction involvement in secretion: the pancreas experience. Clin Exp Pharmacol Physiol 23:1053–1057

Meehan WJ, Samant RS, Hopper JE, Carrozza MJ, Shevde LA, Workman JL, Eckert KA, Verderame MF, Welch DR (2004) Breast cancer metastasis suppressor 1 (BRMS1) forms complexes with retinoblastoma-binding protein 1 (RBP1) and the mSin3 histone deacetylase complex and represses transcription. J Biol Chem 279:1562–1569

Mehta PP, Bertram JS, Loewenstein WR (1986) Growth inhibition of transformed cells correlates with their junctional communication with normal cells. Cell 44:187

Mehta PP, Perez-Stable C, Nadji M, Mian M, Asotra K, Roos BA (1999) Suppression of human prostate cancer cell growth by forced expression of connexin genes. Dev Genet 24:91–110

Mercier F, Hatton GI (2001) Connexin26 and basic fibroblast growth factor are expressed primarily in the subpial and subependymal layers in adult brain parenchyma: roles in stem cell proliferation and morphological plasticity? J Comp Neurol 431:88–104

Mesnil M, Crespin S, Avanzo JL, Zaidan-Dagli ML (2005) Defective gap junctional intercellular communication in the carcinogenic process. Biochim Biophys Acta 1719:125–145

Miller T, Dahl G, Werner R (1988) Structure of a gap junction gene: rat connexin-32. Biosci Rep 8:455–464

Minogue PJ, Tong JJ, Arora A, Russell-Eggitt I, Hunt DM, Moore AT et al (2009) A mutant connexin50 with enhanced hemichannel function leads to cell death. Invest Ophthalmol Vis Sci 50:5837–5845

Mitchell JA, Lye SJ (2005) Differential activation of the connexin43 promoter by dimers of activator protein-1 transcription factors in myometrial cells. Endocrinology 146:2048–2054

Mitchell JA, Ou C, Chen Z, Nishimura T, Lye SJ (2001) Parathyroid hormone-induced up-regulation of connexin-43 messenger ribonucleic acid (mRNA) is mediated by sequences within both the promoter and the 3′untranslated region of the mRNA. Endocrinology 142.907–915

Moennikes O, Buchmann A, Romualdi A, Ott T, Werringloer J, Willecke K, Schwarz M (2000) Lack of phenobarbital-mediated promotion of hepatocarcinogenesis in connexin32-null mice. Cancer Res 60:5087–5091

Moorby C, Patel M (2001) Dual functions for connexins: Cx43 regulates growth independently of gap junction formation. Exp Cell Res 271:238–248

Muller T, Moller T, Neuhaus J, Kettenmann H (1996) Electrical coupling among Bergmann glial cells and its modulation by glutamate receptor activation. Glia 17:274–284

Munhoz Essenfelder G, Bruzzone R, Lamartine J, Charollais A, Blanchet-Bardon C, Barbe MT et al (2004) Connexin30 mutations responsible for hidrotic ectodermal dysplasia cause abnormal hemichannel activity. Hum Mol Genet 13:1703–1714

Musil LS, Goodenough DA (1993) Multisubunit assembly of an integral plasma membrane channel protein, gap junction connexin43, occurs after exit from the ER. Cell 74:1065–1077

Nadarajah B, Parnavelas JG (1999) Gap junction-mediated communication in the developing and adult cerebral cortex. Novartis Found Symp 219:157–170

Nadarajah B, Jones AM, Evans WH, Parnavelas JG (1997) Differential expression of connexins during neocortical development and neuronal circuit formation. J Neurosci 17:3096–3111

Nagasawa K, Chiba H, Fujita H, Kojima T, Saito T, Endo T, Sawada N (2006) Possible involvement of gap junctions in the barrier function of tight junctions of brain and lung endothelial cells. J Cell Physiol 208:123–132

Nagy JI, Li WE (2000) A brain slice model for in vitro analyses of astrocytic gap junction and connexin43 regulation: actions of ischemia, glutamate and elevated potassium. Eur J Neurosci 12:4567–4572

Nagy JI, Rash JE (2000) Connexins and gap junctions of astrocytes and oligodendrocytes in the CNS. Brain Res Brain Res Rev 32:29–44

Nagy JI, Li X, Rempel J, Stelmack G, Patel D, Staines WA et al (2001) Connexin26 in adult rodent central nervous system: demonstration at astrocytic gap junctions and co-localization with con-nexin30 and connexin43. J Comp Neurol 441:302–323

Naus CC (2002) Gap junctions and tumour progression. Can J Physiol Pharmacol 80:136–141

Naus CC, Bechberger JF, Zhang Y, Venance L, Yamasaki H, Juneja SC, Kidder GM, Giaume C (1997) Altered gap junctional communication, intercellular signaling, and growth in cultured astrocytes deficient in connexin43. J Neurosci Res 49:528–540

Naus CC, Bond SL, Bechberger JF, Rushlow W (2000) Identification of genes differentially expressed in C6 glioma cells transfected with connexin43. Brain Res Brain Res Rev 32:259–266

Neijssen J, Herberts C, Drijfhout JW, Reits E, Janssen L, Neefjes J (2005) Cross-presentation by intercellular peptide transfer through gap junctions. Nature 434:83–88

Nelles E, Bützler C, Jung D, Temme A, Gabriel HD, Dahl U, Traub O, Stümpel F, Jungermann K, Zielasek J et al (1996) Defective propagation of signals generated by sympathetic nerve stimu-lation in the liver of connexin32-deficient mice. Proc Natl Acad Sci U S A 93:9565–9570

Nelson PJ, Daniel TO (2002) Emerging targets: molecular mechanisms of cell contact-mediated growth control. Kidney Int 61:99–105

Neuhaus IM, Dahl G, Werner R (1995) Use of alternate promoters for tissue-specific expression of the gene coding for connexin32. Gene 158:257–262

Neuhaus IM, Bone L, Wang S, Ionasescu V, Werner R (1996) The human connexin32 gene is transcribed from two tissue-specific promoters. Biosci Rep 16:239–248

Niessen H, Willecke K (2000) Strongly decreased gap junctional permeability to inositol 1,4,5-trisphosphate in connexin32 deficient hepatocytes. FEBS Lett 466:112–114

Nishimura T, Dunk C, Lu Y, Feng X, Gellhaus A, Winterhager E, Rossant J, Lye SJ (2004) Gap junctions are required for trophoblast proliferation in early human placental development. Placenta 25:595–607

Okano J, Takigawa T, Seki K, Suzuki S, Shiota K, Ishibashi M (2005) Transforming growth factor beta 2 promotes the formation of the mouse cochleovestibular ganglion in organ culture. Int J Dev Biol 49:23–31

Olbina G, Eckhart W (2003) Mutations in the second extracellular region of connexin43 prevent localization to the plasma membrane, but do not affect its ability to suppress cell growth. Mol Cancer Res 1:690–700

Omori Y, Zaidan Dagli ML, Yamakage K, Yamasaki H (2001) Involvement of gap junctions in tumor suppression: analysis of genetically manipulated mice. Mutat Res 477:191–196

Orthmann-Murphy JL, Salsano E, Abrams CK, Bizzi A, Uziel G, Freidin M et al (2009) Hereditary spastic paraplegia is a novel phenotype for GJA12/GJC2 mutations. Brain 132:426–438

Oshima A, Doi T, Mitsuoka K, Maeda S, Fujiyoshi Y (2003) Roles of Met-34, Cys-64, Arg-75 in the assembly of human connexin 26. J Biol Chem 278:1807–1816

Otsuka F, Shimasaki S (2002) A novel function of bone morphogenetic protein-15 in the pituitary: selective synthesis and secretion of FSH by gonadotropes. Endocrinology 143:4938–4941

Oviedo-Orta E, Howard Evans W (2004) Gap junctions and connexin-mediated communication in the immune system. Biochim Biophys Acta 1662:102–112

Oyamada M, Oyamada Y, Takamatsu T (2005) Regulation of connexin expression. Biochim Biophys Acta 1719:6–23

Palka HL, Park M, Tonks NK (2003) Hepatocyte growth factor receptor tyrosine kinase met is a substrate of the receptor proteintyrosine phosphatase DEP-1. J Biol Chem 278:5728–5735

Palmer JW, Tandler B, Hoppel CL (1977) Biochemical properties of subsarcolemmal and interfibrillar mitochondria isolated from rat cardiac muscle. J Biol Chem 252:8731–8739

Paznekas WA, Karczeski B, Vermeer S, Lowry RB, Delatycki M, Laurence F et al (2009) GJA1 mutations, variants, and connexin 43 dysfunction as it relates to the oculodentodigital dysplasia phenotype. Hum Mutat 30:724–733

Penes MC, Li X, Nagy JI (2005) Expression of zonula occludens-1 (ZO-1) and the transcription factor ZO-1-associated nucleic acidbinding protein (ZONAB)-MsY3 in glial cells and colocalization at oligodendrocyte and astrocyte gap junctions in mouse brain. Eur J Neurosci 22:404–418

Perbal B (2003) The CCN3 (NOV) cell growth regulator: a new tool for molecular medicine. Expert Rev Mol Diagn 3:597–604

Petrich BG, Gong X, Lerner DL, Wang X, Brown JH, Saffitz JE, Wang Y (2002) c-Jun N-terminal kinase activation mediates downregulation of connexin43 in cardiomyocytes. Circ Res 91:640–647

Petrich BG, Eloff BC, Lerner DL, Kovacs A, Saffitz JE, Rosenbaum DS, Wang Y (2004) Targeted activation of c-Jun N-terminal kinase in vivo induces restrictive cardiomyopathy and conduction defects. J Biol Chem 279:15330–15338

Pfeifer I, Anderson C, Werner R, Oltra E (2004) Redefining the structure of the mouse connexin43 gene: selective promoter usage and alternative splicing mechanisms yield transcripts with different translational efficiencies. Nucleic Acids Res 32:4550–4562

Piechocki MP, Burk RD, Ruch RJ (1999) Regulation of connexin32 and connexin43 gene expression by DNA methylation in rat liver cells. Carcinogenesis 20:401–406

Pozas E, Ibanez CF (2005) GDNF and GFRalpha1 promote differentiation and tangential migration of cortical GABAergic neurons. Neuron 45:701–713

Pyronnet S, Pradayrol L, Sonenberg N (2000) A cell cycle-dependent internal ribosome entry site. Mol Cell 5:607–616

Qin H, Shao Q, Igdoura SA, Alaoui-Jamali MA, Laird DW (2003) Lysosomal and proteasomal degradation play distinct roles in the life cycle of Cx43 in gap junctional intercellular communication deficient and -competent breast tumor cells. J Biol Chem 278:30005–30014

Reaume AG, de Sousa PA, Kulkarni S, Langille BL, Zhu D, Davies TC, Juneja SC, Kidder GM, Rossant J (1995) Cardiac malformation in neonatal mice lacking connexin43. Science 267:1831–1834

Ressot C, Gome's D, Dautigny A, Pham-Dinh D, Bruzzone R (1998) Connexin32 mutations associated with X-linked Charcot-Marie-Tooth disease show two distinct behaviors: loss of function and altered gating properties. J Neurosci 18:4063–4075

Retamal MA, Cortes CJ, Reuss L, Bennett MV, Saez JC (2006) S-nitrosylation and permeation through connexin 43 hemichannels in astrocytes: induction by oxidant stress and reversal byreducing agents. Proc Natl Acad Sci U S A 103:4475–4480

Richard G (2000) Connexins: a connection with the skin. Exp Dermatol 9:77–96

Rousset B (1996) Introduction to the structure and functions of junction communications or gap junctions. Ann Endocrinol (Paris) 57:476–480

Roy H, Bhardwaj S, Yla-Herttuala S (2006) Biology of vascular endothelial growth factors. FEBS Lett 580:2879–2887

Saez JC, Connor JA, Spray DC, Bennett MV (1989) Hepatocyte gap junctions are permeable to the second messenger, inositol 1,4,5-trisphosphate, and to calcium ions. Proc Natl Acad Sci U S A 86:2708–2712

Saez JC, Berthoud VM, Branes MC, Martinez AD, Beyer EC (2003) Plasma membrane channels formed by connexins: their regulation and functions. Physiol Rev 83:1359–1400

Saez JC, Retamal MA, Basilio D, Bukauskas FF, Bennett MV (2005) Connexin-based gap junction hemichannels: gating mechanisms. Biochim Biophys Acta 1711:215–224

Saunders MM, Seraj MJ, Li Z, Zhou Z, Winter CR, Welch DR, Donahue HJ (2001) Breast cancer metastatic potential correlates with a breakdown in homospecific and heterospecific gap junctional intercellular communication. Cancer Res 61:1765–1767

Segretain D, Falk MM (2004) Regulation of connexin biosynthesis, assembly, gap junction formation, and removal. Biochim Biophys Acta 1662:3–21

Serre-Beinier V, Le Gurun S, Belluardo N, Trovato-Salinaro A, Charollais A, Haefliger JA, Condorelli DF, Meda P (2000) Cx36 preferentially connects β-cells within pancreatic islets. Diabetes 49:727–734

Seul KH, Beyer EC (2000) Mouse connexin37: gene structure and promoter analysis. Biochim Biophys Acta 1492:499–504

Seul KH, Tadros PN, Beyer EC (1997) Mouse connexin40: gene structure and promoter analysis. Genomics 46:120–126

Severs NJ, Dupont E, Coppen SR, Halliday D, Inett E, Baylis D, Rothery S (2004) Remodelling of gap junctions and connexin expression in heart disease. Biochim Biophys Acta 1662:138–148

Shao Q, Wang H, McLachlan E, Veitch GI, Laird DW (2005) Down-regulation of Cx43 by retroviral delivery of small interfering RNA promotes an aggressive breast cancer cell phenotype. Cancer Res 65:2705–2711

Shaw RM, Fay AJ, Puthenveedu MA, von Zastrow M, Jan YN, Jan LY (2007) Microtubule plus-end-tracking proteins target gap junctions directly from the cell interior to adherens junctions. Cell 128:547–560

Sheikh F, Hirst CJ, Jin Y, Bock ME, Fandrich RR, Nickel BE, Doble BW, Kardami E, Cattini PA (2004) Inhibition of TGFbeta signaling potentiates the FGF-2-induced stimulation of cardiomyocyte DNA synthesis. Cardiovasc Res 64:516–525

Shestopalov VI, Panchin Y (2008) Pannexins and gap junction protein diversity. Cell Mol Life Sci 65:376–394

Shields CR, Klooster C, Claassen Y, Hussain MU, Zoidl G, Dermietzel R, Kamermans M (2007) Retinal horizontal cell-specific promoter activity and protein expression of zebrafish connexin 52.6 and connexin 55.5. J Comp Neurol 501:765–779

Simon AM, Goodenough DA, Li E, Paul DL (1997) Female infertility in mice lacking connexin 37. Nature (Lond) 385:525–529

Simon AM, Goodenough DA, Paul DL (1998) Mice lacking connexin40 have cardiac conduction abnormalities characteristic of atrioventricular block and bundle branch block. Curr Biol 8:295–298

Singh D, Lampe PD (2003) Identification of connexin-43 interacting proteins. Cell Commun Adhes 10:215–220

Singh D, Solan JL, Taffet SM, Javier R, Lampe PD (2005) Connexin 43 interacts with zona occludens-1 and -2 proteins in a cell cycle stage-specific manner. J Biol Chem 280:30416–30421

SiuYi Leung D, Unsicker K, Reuss B (2001) Gap junctions modulate survival-promoting effects of fibroblast growth factor-2 on cultured midbrain dopaminergic neurons. Mol Cell Neurosci 18:44–55

Söhl G, Willecke K (2003) An update on connexin genes and their nomenclature in mouse and man. Cell Commun Adhes 10:173–180

Söhl G, Willecke K (2004) Gap junctions and the connexin protein family. Cardiovasc Res 62:228–232

Söhl G, Gillen C, Bosse F, Gleichmann M, Müller HW, Willecke K (1996) A second alternative transcript of the gap junction gene connexin32 is expressed in murine Schwann cells and modulated in injured sciatic nerve. Eur J Cell Biol 69:267–275

Söhl G, Degen J, Teubner B, Willecke K (1998) The murine gap junction gene connexin36 is highly expressed in mouse retina and regulated during brain development. FEBS Lett 428:27–31

Söhl G, Eiberger J, Jung YT, Kozak CA, Willecke K (2001) The mouse gap junction gene connexin29 is highly expressed in sciatic nerve and regulated during brain development. Biol Chem 382:973–978

Solan JL, Lampe PD (2005) Connexin phosphorylation as a regulatory event linked to gap junction channel assembly. Biochim Biophys Acta 1711:154–163

Sosinsky GE, Nicholson BJ (2005) Structural organization of gap junction channels. Biochim Biophys Acta 1711:99–125

Srisakuldee W, Nickel BE, Fandrich RR, Jiang ZS, Kardami E (2006) Administration of FGF-2 to the heart stimulates connexin-43 phosphorylation at protein kinase C target sites. Cell Commun Adhes 13:13–19

Stains JP, Civitelli R (2005) Gap junctions regulate extracellular signal-regulated kinase signaling to affect gene transcription. Mol Biol Cell 16:64–72

Stains JP, Lecanda F, Screen J, Towler DA, Civitelli R (2003) Gap junctional communication modulates gene transcription by altering the recruitment of Sp1 and Sp3 to connexin-response elements in osteoblast promoters. J Biol Chem 278:24377–24387

Stock A, Sies H (2000) Thyroid hormone receptors bind to an element in the connexin43 promoter. Biol Chem 381:973–979

Talhouk RS, Elble RC, Bassam R, Daher M, Sfeir A, Mosleh LA, El-Khoury H, Hamoui S, Pauli BU, El-Sabban ME (2005) Developmental expression patterns and regulation of connexins in the mouse mammary gland: expression of connexin30 in lactogenesis. Cell Tissue Res 319:49–59

Tamaddon HS, Vaidya D, Simon AM, Paul DL, Jalife J, Morley GE (2000) High-resolution optical mapping of the right bundle branch in connexin40 knockout mice reveals slow conduction in the specialized conduction system. Circ Res 87:929–936

Tamamori-Adachi M, Hayashida K, Nobori K, Omizu C, Yamada K, Sakamoto N, Kamura T, Fukuda K, Ogawa S, Nakayama KI, Kitajima S (2004) Down-regulation of p27Kip1 promotes cell proliferation of rat neonatal cardiomyocytes inducedby nuclear expression of cyclin D1 and CDK4. Evidence for impaired Skp2-dependent degradation of p27 in terminal differentiation. J Biol Chem 279:50429–50436

Temme A, Buchmann A, Gabriel HD, Nelles E, Schwarz M, Willecke K (1997) High incidence of spontaneous and chemically induced liver tumors in mice deficient for connexin32. Curr Biol 7:713–716

Teubner B, Odermatt B, Güldenagel M, Söhl G, Degen J, Bukauskas F, Kronengold J, Verselis VK, Jung YT, Kozak CA et al (2001) Functional expression of the new gap junction gene connexin47 transcribed in mouse brain and spinal cord neurons. J Neurosci 21:1117–1126

Teunissen BE, Jansen AT, van Amersfoorth SC, O'Brien TX, Jongsma HJ, Bierhuizen MF (2003) Analysis of the rat connexin 43 proximal promoter in neonatal cardiomyocytes. Gene 322:123–136

Toyofuku T, Akamatsu Y, Zhang H, Kuzuya T, Tada M, Hori M (2001) c-Src regulates the interaction between connexin-43 and ZO-1 in cardiac myocytes. J Biol Chem 276:1780–1788

Trexler EB, Bennett MV, Bargiello TA, Verselis VK (1996) Voltage gating and permeation in a gap junction hemichannel. Proc Natl Acad Sci U S A 93:5836–5841

Trosko JE, Ruch RJ (2002) Gap junctions as targets for cancer chemoprevention and chemotherapy. Curr Drug Targets 3:465–482

Turek-Plewa J, Jagodziński PP (2005) The role of mammalian DNA methyltransferases in the regulation of gene expression. Cell Mol Biol Lett 10:631–647

Unwin PN, Zampighi G (1980) Structure of the junction between communicating cells. Nature 283:545–549

Ursitti JA, Petrich BG, Lee PC, Resneck WG, Ye X, Yang J, Randall WR, Bloch RJ, Wang Y (2007) Role of an alternatively spliced form of alphaII-spectrin in localization of connexin 43 in cardiomyocytes and regulation by stress-activated protein kinase. J Mol Cell Cardiol 42:572–581

Vaidya D, Tamaddon HS, Lo CW, Taffet SM, Delmar M, Morley GE, Jalife J (2001) Null mutation of connexin43 causes slow propagation of ventricular activation in the late stages of mouse embryonic development. Circ Res 88:1196–1202

Valiunas V, Polosina YY, Miller H, Potapova IA, Valiuniene L, Doronin S, Mathias RT, Robinson RB, Rosen MR, Cohen IS, Brink PR (2005) Connexin-specific cell-to-cell transfer of short interfering RNA by gap junctions. J Physiol 568:459–468

Van der Heyden MA, Rook MB, Hermans MM, Rijksen G, Boonstra J, Defize LH, Destree OH (1998) Identification of connexin43 as a functional target for Wnt signalling. J Cell Sci 111:1741–1749

Van Es RJ, Wittebol-Post D, Beemer FA (2007) Oculodentodigital dysplasia with mandibular retrognathism and absence of syndactyly: a case report with a novel mutation in the connexin 43 gene. Int J Oral Maxillofac Surg 36:858–860

Van Rijen HV, van Kempen MJ, Postma S, Jongsma HJ (1998) Tumour necrosis factor alpha alters the expression of connexin43, connexin40, and connexin37 in human umbilical vein endothelial cells. Cytokine 10:258–264

Vinken M, Vanhaecke T, Papeleu P, Snykers S, Henkens T, Rogiers V (2006) Connexins and their channels in cell growth and cell death. Cell Signal 18:592–600

Vinken M, De Rop E, Decrock E, De Vuyst E, Leybaert L, Vanhaecke T, Rogiers V (2009) Epigenetic regulation of gap junctional intercellular communication: more than a way to keep cells quiet? Biochim Biophys Acta 1795:53–61

Warn-Cramer BJ, Lampe PD, Kurata WE, Kanemitsu MY, Loo LW, Eckhart W, Lau AF (1996) Characterization of the mitogen-activated protein kinase phosphorylation sites on the connexin-43 gap junction protein. J Biol Chem 271:3779–3786

Wei CJ, Xu X, Lo CW (2004) Connexins and cell signaling in development and disease. Annu Rev Cell Dev Biol 20:811–838

Wei CJ, Francis R, Xu X, Lo CW (2005) Connexin43 associated with an N-cadherin-containing multiprotein complex is required for gap junction formation in NIH3T3 cells. J Biol Chem 280:19925–19936

Weissman TA, Requelme PA, Ivic L, Flint AC, Kriegstein AR (2004) Calcium waves propagate through radial glial cells and modulate proliferation in the developing neocortex. Neuron 43:647–661

Weng S, Lauven M, Schaefer T, Polontchouk L, Grover R, Dhein S (2002) Pharmacological modification of gap junction coupling by an antiarrhythmic peptide via protein kinase C activation. FASEB J 16:1114–1116

Werner R (2000) IRES elements in connexin genes: a hypothesis explaining the need for connexins to be regulated at the translational level. IUBMB Life 50:173–176

White TW, Paul DL (1999) Genetic diseases and gene knockouts reveal diverse connexin functions. Annu Rev Physiol 61:283–310

White TW, Bruzzone R, Paul DL (1995) The connexin family of intercellular channel forming proteins. Kidney Int 48:1148–1157

Willecke K, Eiberger J, Degen J, Eckardt D, Romualdi A, Guldenagel M, Deutsch U, Sohl G (2002) Structural and functional diversity of connexin genes in the mouse and human genome. Biol Chem 383:725–737

Wong CW, Christen T, Roth I, Chadjichristos CE, Derouette JP, Foglia BF et al (2006) Connexin37 protects against atherosclerosis by regulating monocyte adhesion. Nat Med 12:950–954

Wu JC, Tsai RY, Chung TH (2003) Role of catenins in the development of gap junctions in rat cardiomyocytes. J Cell Biochem 88:823–835

Wu ZH, Mabb A, Miyamoto S (2005) PIDD: a switch hitter. Cell 123:980–982

Xia X, Batra N, Shi Q, Bonewald LF, Sprague E, Jiang JX (2010) Prosglandin promotion of osteocyte gap junction function through transcriptional regulation of connexin 43 by glycogen synthase kinase 3/_-catenin signaling. Mol Cell Biol 30:206–219

Xu X, Li WE, Huang GY, Meyer R, Chen T, Luo Y, Thomas MP, Radice GL, Lo CW (2001) N-cadherin and Cx43 alpha1 gap junctions modulates mouse neural crest cell motility via distinct pathways. Cell Commun Adhes 8:321–324

Ya J, Erdtsieck-Ernste EB, de Boer PA, van Kempen MJ, Jongsma H, Gros D, Moorman AF, Lamers WH (1998) Heart defects in connexin43-deficient mice. Circ Res 82:360–366

Yang B, Lin H, Xiao J, Lu Y, Luo X, Li B, Zhang Y, Xu C, Bai Y, Wang H, Chen G, Wang Z (2007) The muscle-specific microRNA miR-1 regulates cardiac arrhythmogenic potential by targeting GJA1 and KCNJ2. Nat Med 13:486–491

Yang N, Coukos G, Zhang L (2008) MicroRNA epigenetic alterations in human cancer: one step forward in diagnosis and treatment. Int J Cancer 122:963–968

Yeager M, Unger VM, Falk MM (1998) Synthesis, assembly and structure of gap junction intercellular channels. Curr Opin Struct Biol 8:517–524

Yi ZC, Wang H, Zhang GY, Xia B (2007) Downregulation of connexin 43 in nasopharyngeal carcinoma cells is related to promoter methylation. Oral Oncol 43:898–904

Yu W, Dahl G, Werner R (1994) The connexin43 gene is responsive to oestrogen. Proc R Soc Lond B Biol Sci 255

Zhang JT, Chen M, Foote CI, Nicholson BJ (1996) Membrane integration of in vitro-translated gap junctional proteins: Co- and posttranslational mechanisms. Mol Biol Cell 7:471–482

Zhang YW, Kaneda M, Morita I (2003a) The gap junction-independent tumor-suppressing effect of connexin43. J Biol Chem 278:44852–44856

Zhang YW, Nakayama K, Nakayama K, Morita I (2003b) A novel route for connexin43 to inhibit cell proliferation: negative regulation of S-phase kinase-associated protein (Skp 2). Cancer Res 63:1623–1630

Zhou L, Kasperek EM, Nicholson BJ (1999) Dissection of the molecular basis of pp 60(v-src) induced gating of connexin 43 gap junction channels. J Cell Biol 144:1033–1045